Thinking Critically: Climate Change

Andrea C. Nakaya

San Diego, CA

About the Author

Andrea C. Nakaya, a native of New Zealand, holds a bachelor's degree in English and a master of arts degree in communications from San Diego State University. She has written and edited numerous books on current issues. She currently lives in Encinitas, California, with her husband and their two children, Natalie and Shane.

Printed in the United States

For more information, contact:
ReferencePoint Press, Inc.
PO Box 27779
San Diego, CA 92198
www. ReferencePointPress.com

Picture Credits:
NASA/Science Photo Library: 8
Maury Aaseng: 15, 20, 29, 34, 38, 45, 53, 58

LIBRARY OF CONGRESS CATALOGING-IN-PUBLICATION DATA

Nakaya, Andrea C., 1976– author.
Climate change / by Andrea C. Nakaya.
pages cm. -- (Thinking critically series)
Includes bibliographical references and index.
Audience: Grade 9 to 12.
ISBN-13: 978-1-60152-732-5 (hardback : alk. paper)
ISBN-10: 1-60152-732-2 (hardback : alk. paper)
1. Climatic changes--Effect of human beings on--Popular works. I. Title.
GF47.N35 2015
363.738'74--dc23

2014005802

Contents

Foreword 4
Overview: Climate Change 6

Chapter One: Is Climate Change Caused by Human Activity?
The Debate at a Glance 12
Climate Change Is Caused by Human Activity 13
Climate Change Is Not Caused by Human Activity 19

Chapter Two: Has the Extent of Climate Change Been Overstated?
The Debate at a Glance 24
The Extent of Climate Change Has Been Overstated 25
The Extent of Climate Change Has Not Been Overstated 31

Chapter Three: Will Climate Change Be Harmful to Society?
The Debate at a Glance 36
Climate Change Will Be Harmful to Society 37
Climate Change Will Not Be Harmful to Society 43

Chapter Four: Should Society Try to Reduce Climate Change?
The Debate at a Glance 49
Society Should Try to Reduce Climate Change 50
Society Should Not Try to Reduce Climate Change 56

Source Notes 62
Climate Change Facts 68
Related Organizations and Websites 71
For Further Research 75
Index 77

Foreword

"Literacy is the most basic currency of the knowledge economy we're living in today." Barack Obama (at the time a senator from Illinois) spoke these words during a 2005 speech before the American Library Association. One question raised by this statement is: What does it mean to be a literate person in the twenty-first century?

E.D. Hirsch Jr., author of *Cultural Literacy: What Every American Needs to Know*, answers the question this way: "To be culturally literate is to possess the basic information needed to thrive in the modern world. The breadth of the information is great, extending over the major domains of human activity from sports to science."

But literacy in the twenty-first century goes beyond the accumulation of knowledge gained through study and experience and expanded over time. Now more than ever literacy requires the ability to sift through and evaluate vast amounts of information and, as the authors of the Common Core State Standards state, to "demonstrate the cogent reasoning and use of evidence that is essential to both private deliberation and responsible citizenship in a democratic republic."

The Thinking Critically series challenges students to become discerning readers, to think independently, and to engage and develop their skills as critical thinkers. Through a narrative-driven, pro/con format, the series introduces students to the complex issues that dominate public discourse—topics such as gun control and violence, social networking, and medical marijuana. All chapters revolve around a single, pointed question such as Can Stronger Gun Control Measures Prevent Mass Shootings?, or Does Social Networking Benefit Society?, or Should Medical Marijuana Be Legalized? This inquiry-based approach introduces student researchers to core issues and concerns on a given topic. Each chapter includes one part that argues the affirmative and one part that argues the negative—all written by a single author. With the single-author format the predominant arguments for and against an issue can be synthesized into clear, accessible discussions supported by details and evidence including relevant facts, direct quotes, current examples, and statistical illustrations. All volumes include focus questions to guide students as they read each pro/con discussion, a list of key facts, and an annotated list of related organizations and websites for conducting further research.

The authors of the Common Core State Standards have set out the particular qualities that a literate person in the twenty-first century must have. These include the ability to think independently, establish a base of knowledge across a wide range of subjects, engage in open-minded but discerning reading and listening, know how to use and evaluate evidence, and appreciate and understand diverse perspectives. The new Thinking Critically series supports these goals by providing a solid introduction to the study of pro/con issues.

Climate Change

The topic of climate change received intense public attention with the 2006 release of the documentary *An Inconvenient Truth*. This Academy Award–winning film documents the efforts of former US vice president Al Gore to educate the public about climate change. Gore warns that human action is causing significant changes in the climate and that if humans do not attempt to stop what is happening the future impacts will be severe. For example, he warns that global sea levels will rise dramatically, storms will become increasingly intense and destructive, and many species of plants and animals will be driven to extinction. He says, "What we take for granted might not be here for our children."[1]

Years later, Gore continues his mission to educate society about climate change. He is not alone. A large number of scientists agree that the world's climate is changing. However, there is less agreement on what the effects of that change will be and whether society should—or even can—try to do anything to address it. The answers to these questions have extremely important implications. Any attempts to address climate change will necessitate major social changes that will definitely be inconvenient, as the title of Gore's film points out. However, if his warnings are true, problems such as rising sea level and species extinction will be more than inconvenient; they could result in the deaths of millions of people around the globe.

How the Climate Changes

Earth's climate is determined by a complex interaction between how much energy it receives from the sun and how much it releases back into space. Earth warms when it receives energy from the sun but cools when it releases energy. The balance between these two is what determines the climate, and that balance is ever changing. Numerous factors can change how much energy is absorbed and released. For example, there are variations in solar energy over time. There are also variations in the reflectivity of Earth; for instance, ice reflects the sun's energy back into space, so when Earth has more areas of ice, more energy is released.

By examining evidence of the past—which can be found in ice cores, tree rings, and ocean sediments—scientists have discovered that climate has changed many times throughout history. There have been periods when most of the planet was covered in ice and periods when the weather was much warmer. At present Earth's climate is in a period of warming. The Environmental Protection Agency (EPA) explains that there is widespread evidence of warming from both temperature records and observations of the natural landscape. It says, "Multiple temperature records from all over the world have all shown a warming trend, and these records have been deemed reliable by the National Aeronautics and Space Administration (NASA), and the National Oceanic and Atmospheric Administration (NOAA), among others." In addition, it says, "Other observations that point to higher global temperature includes: warmer oceans, melting arctic sea ice and glaciers, sea level rise, increasing precipitation, and changing wind patterns."[2]

While the overall long-term trend is that temperatures are increasing, the changing climate is also causing unusual and extreme weather events around the world. These events include heat waves but also periods of record snowfall or extreme cold such as occurred in parts of the United States in the winter of 2013–2014.

Greenhouse Gases

Most scientists believe that one of the main causes of warming is the greenhouse effect. This occurs when gases in the atmosphere trap the sun's heat so that it is not reflected back into space, similar to how a greenhouse traps heat inside. The most common greenhouse gases are carbon dioxide, methane, and nitrous oxide. Data show that atmospheric levels of greenhouse gases have been steadily increasing since the Industrial Revolution in the

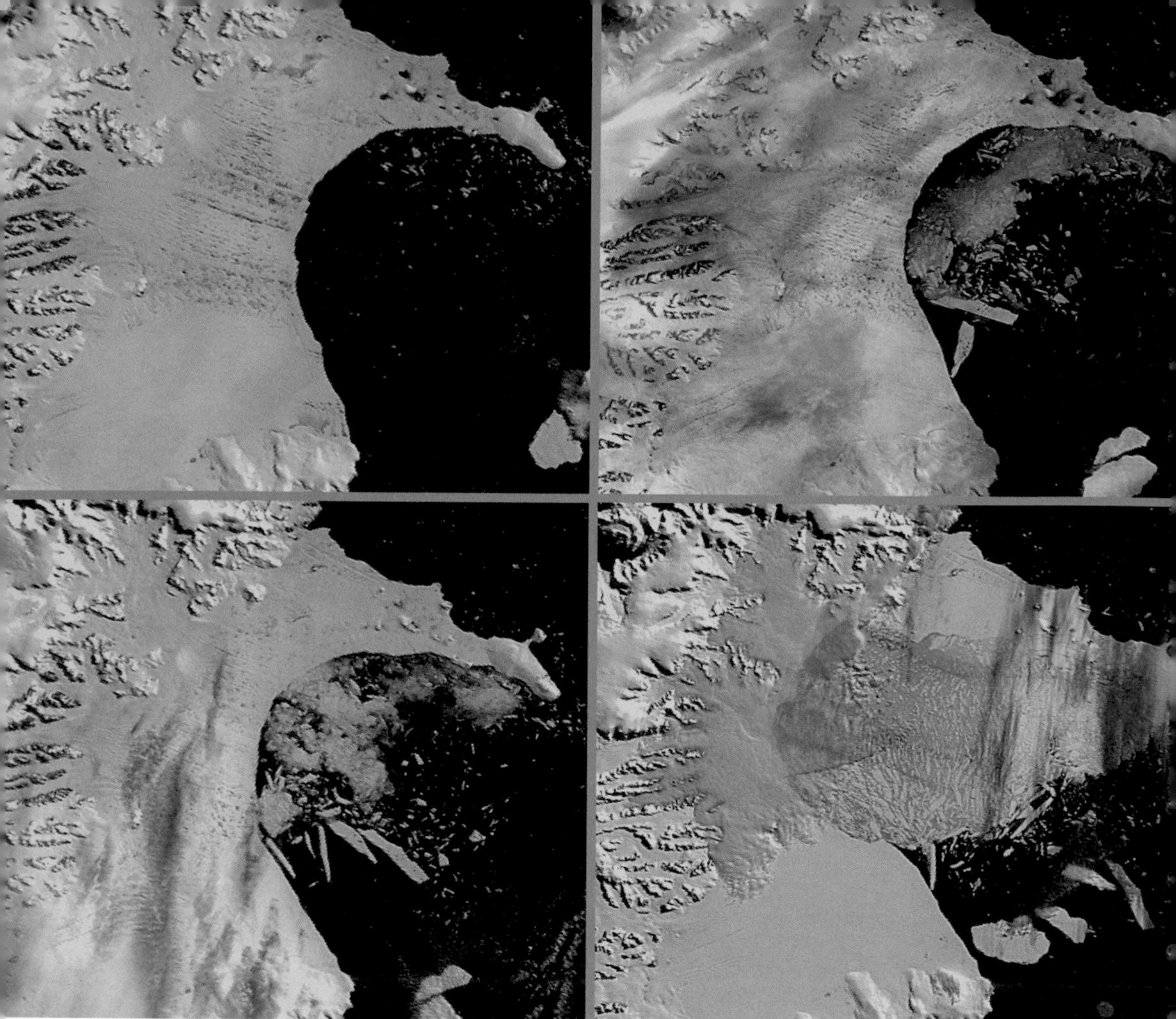

Evidence of climate change can be clearly seen, according to scientists, in NASA satellite images of a collapsing Antarctic ice shelf. The four images above show the Larsen B Ice Shelf collapsing into the Weddell Sea in 2002. Top left, an image taken on January 31 shows dark blue melt ponds in the ice shelf. Top right, an image taken on February 17 shows an area of about 309 square miles (800 sq km) that has broken off, leaving icebergs in the sea and melt ponds draining through new cracks within the ice shelf. Bottom left, an image taken on February 23 shows continued collapse of the ice shelf. Bottom right, an image taken on March 5 shows thousands of icebergs including a large area of small icebergs (light blue), also known as bergy bits.

late eighteenth century, and at the same time Earth's temperature has been increasing.

While natural events such as volcanic eruptions can change the amount of greenhouse gases in Earth's atmosphere, human activity also has a significant effect. Humans create large amounts of greenhouse gases when they use fossil fuels to create energy. The world greatly depends on these fuels. Even though alternate fuels have been developed in recent years, fossil fuels still account for the majority of the global energy supply. According to the International Energy Agency, in 2011 they supplied about 82 percent of the world's energy.

Some factors that affect climate, such as greenhouse gases or how much solar energy Earth receives, are well understood. But other factors, known as feedback effects, make it difficult to predict how climate will change in the future. Feedback effects either amplify or diminish the way greenhouse gases and other factors affect the climate. The most important feedback effects are clouds, sea ice, and water vapor. There is a large amount of uncertainty about exactly how much of an effect they have. For instance, clouds can either increase warming by trapping heat or decrease it by reflecting sunlight away from Earth's surface. The effect depends on the type and the altitude of the cloud.

Climate Affects Everything

Changes in climate have the potential to affect everything on Earth. Journalism and research organization Climate Central explains that Earth's land, oceans, plants, and animals are all connected, and a change in one of them means a change in others. It says, "The result of a rise in temperature isn't just a slightly different version of the world we have now." Instead, a changing climate can mean a very different world. For example, climate change is causing the world's glaciers to melt, and scientists predict that this will change the planet's water cycles. If water cycles change, both landscapes and weather could be altered. Climate Central explains, "Mountain glaciers will start to melt, so the rivers they feed will flow differently. More water will evaporate from both land and oceans, leading to more rain and snow in some places and more drought in others. As the oceans warm, major currents may speed up or slow down—which would affect weather patterns."[3] These changes could in turn dramatically affect ecosystems.

In the face of such potentially serious consequences, numerous national and international research efforts have been undertaken to assess the extent of climate change and its possible impacts. The most well-recognized and comprehensive research is that of the United Nations International Panel on Climate Change (IPCC). Established in 1988, the IPCC's mission is to collect and assess information about climate change. Thousands of scientists and other experts contribute to this research, and the organization releases reports, the most recent in 2013. In that report the IPCC states that climate change is undeniably occurring. It says, "The atmosphere and ocean have warmed, the amounts of snow and ice have diminished, sea level has risen, and the concentrations of greenhouse gases have increased." It states, "Each of the last three decades has been successively warmer at the Earth's surface than any preceding decade since 1850."[4] The IPCC also finds it extremely likely that human action is the cause of this warming.

Action to Address Climate Change

While a large number of people agree that climate change is occurring, far fewer agree about whether society should take action to address this change and what exactly should be done. Numerous countries have taken action to reduce their greenhouse gas emissions, since this is widely believed to be the main cause of warming. For example, in 2005 an international treaty called the Kyoto Protocol went into effect. The treaty, ratified by 192 countries, involves agreeing to legally binding emissions limits. However, more than one hundred developing countries, including China and India, are exempt from the treaty, and these countries account for a large percentage of world emissions. The United States, which also causes a significant percentage of world emissions, dropped out of the treaty in 2001.

In addition to such efforts to reduce emissions, it is widely argued that some climate change effects are inevitable and that society should start researching and preparing to adapt to a future in which the climate will be substantially different. According to the American Chemical Society, "The current levels of long-lived atmospheric greenhouse gases and the levels of increased CO_2 and heat absorbed by the world's oceans ensure that the climate will almost certainly continue to increase for decades, even if greenhouse gas and absorbing particle emissions are scaled back to more sustainable levels." As a result, it says, "The world must be prepared to adapt to changes in water supplies, agricultural productivity, severe weather patterns, sea-level rise, ocean acidification and ecosystem viabilities."[5]

An Uncertain Future

Nobody knows for sure how climate change will affect society in the future. While large climactic changes have occurred in the past, and humankind has survived, today Earth has many more people than ever before and depends on large cities and elaborate systems of agriculture, manufacturing, and trade. Climate Central says, "It's one thing for a small band of people to pack up camp and move. . . . It's a very different thing to try to move a city like Cairo or New York or Shanghai because the sea level is rising." It says, "Our civilization, in short, is very highly adapted to the present climate."[6] This means that answers to questions about the future of the climate are extremely important to everyone on Earth.

Chapter One

Is Climate Change Caused by Human Activity?

Climate Change Is Caused by Human Activity

- Deforestation is contributing to climate change.
- The majority of scientists agree that climate change is human-caused.
- Natural processes do not explain the recent period of climate change.
- Human activity is creating so much carbon dioxide that the earth is warming.

Climate Change Is Not Caused by Human Activity

- History shows that Earth's climate always changes over time.
- Changes in the climate are due to natural forces.
- Solar variation explains some of the changes in Earth's climate.
- There is no proof that greenhouse gases cause the temperature to rise.

Climate Change Is Caused by Human Activity

"Scientists have closed the case: Human activity is causing the Earth to get hotter."

—The Environmental Defense Fund is an organization that works to preserve the natural environment.

Environmental Defense Fund, "How Are Humans Responsible for Global Warming?," 2013. www.edf.org.

Consider these questions as you read:

1. Do you agree with the argument that climate change is not the result of natural processes? Why or why not?
2. How persuasive is the argument that humans are upsetting the natural balance of carbon in the world? Explain.
3. How persuasive is the argument that climate change is caused by human activity? Which arguments provide the strongest support for this perspective, and why?

Editor's note: The discussion that follows presents common arguments made in support of this perspective, reinforced by facts, quotes, and examples taken from various sources.

According to a 2012 report by the Food and Agriculture Organization of the United Nations (FAO), forests cover about 31 percent of Earth's land surface. However, every year this percentage grows smaller as humans destroy large areas of forest for logging or to clear land for other uses such as agriculture and residential and commercial development. According to the World Wildlife Federation, about 46 million to 58 million square miles of forest are lost every year. This deforestation is having a significant impact on the climate. Trees capture and store carbon dioxide from the atmosphere, so when there are fewer trees less carbon dioxide is captured and carbon levels increase. In addition, deforestation adds carbon dioxide to the atmosphere because when a tree is cut down it releases its store of carbon dioxide. Associate editor of the *Futurist*, Rick Docksai, says that deforestation currently accounts for 10 to 15 percent of global greenhouse gas emissions—not an insignificant amount. It rivals another well-known source of pollution, as Docksai explains: "As a comparison, guess how much carbon-dioxide generation traceable to human activity comes just from automobile traffic. According to the World Resources Institute, it's around 15%."[7] Human activity is having a significant impact on the world's climate. Deforestation is just one of the many ways this is happening.

Scientific Consensus

Most scientific experts recognize that human activity is causing climate change. Many leading scientific organizations including the World Health Organization, the US National Aeronautics and Space Administration (NASA), and the World Meteorological Organization have issued public statements endorsing this position. The IPCC compiles and assesses information about climate change from thousands of scientists and other experts. In its 2013 report it concludes, "It is *extremely likely* that human influence has been the dominant cause of the observed warming since the mid-20th century."[8]

"It is *extremely likely* that human influence has been the dominant cause of the observed warming since the mid-20th century."[8]

—The Intergovernmental Panel on Climate Change, an international organization that assesses information about climate change and issues reports for the public and policy makers.

Among the general public there is also widespread agreement that human activity is changing the climate. Gallup polls conducted between 2003 and 2013 show that most Americans continue to believe that changes in Earth's temperature are due to human activities. In 2013 that percentage was at 57 percent. This position is on such firm scientific footing that in 2013 the *Los Angeles Times* (a major national newspaper) stated that it would no longer print letters denying that climate change is caused by human activity. The paper's editors noted that such statements are untrue. It says, "Letters that have an untrue basis (for example, ones that say there's no sign humans have caused climate change) do not get printed."[9]

Human-Caused Carbon Emissions Are Driving Climate Change

Most scientists worldwide agree that human activities, specifically those that result in carbon dioxide emissions, are a primary cause of climate change. Levels of CO_2 have risen and fallen in a pattern for hundreds of thousands of years until the 1880s and the Industrial Revolution, when humans started to create high levels of CO_2 emissions. Since then, the level of atmospheric carbon dioxide has rapidly and steadily increased, exceeding 300 parts per million (ppm) for the first time in 1950. By July 2013, that level was approaching 400 ppm, with no sign of slowing.

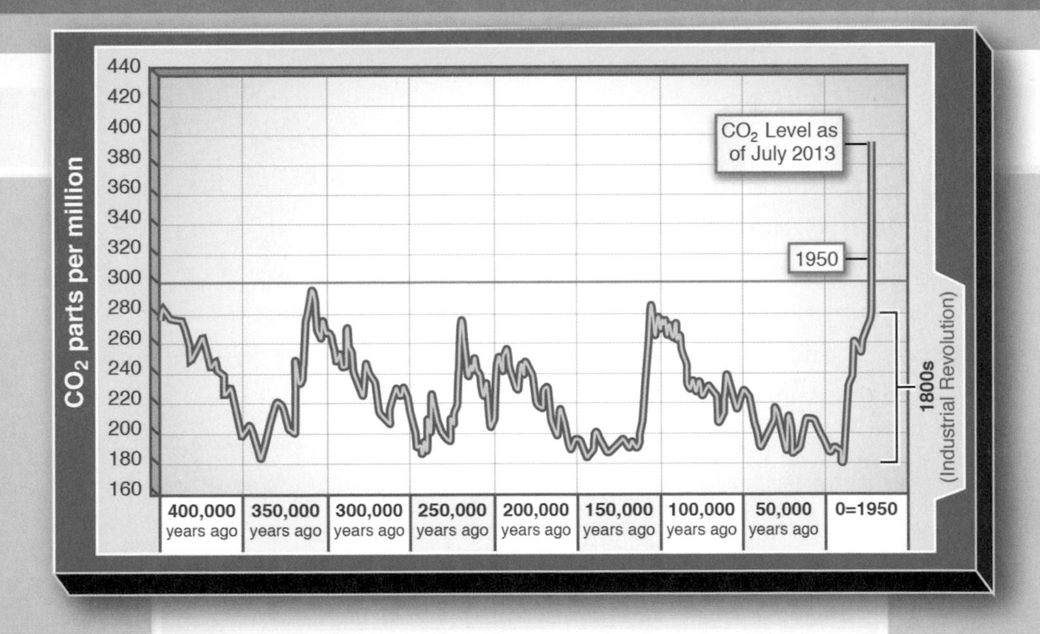

Source: National Aeronautics and Space Administration, Climate Change, "Evidence: Climate Change; How Do We Know?," 2013. http://climate.nasa.gov.

Not a Result of Natural Processes

Earth has undergone numerous cycles of climate change as a result of natural processes. An examination of the fossil and geologic record shows that these were due to the changing positions of the continents, changing patterns of ocean circulation, and changes in Earth's orbit. For example, many geologists believe that in the mid-Cretaceous period—approximately 100 million years ago—Alaska had subtropical conditions, and there were no polar icecaps.

However, evidence clearly shows that the current period of warming is different from these previous cycles. The climate changes caused by natural forces were gradual and occurred over hundreds or thousands of years. This differs from the current warming, which is occurring far more quickly. The Geological Society of America (GSA) examines the impact of natural forces such as changing ocean currents and finds that, "[These] are far too slow to have played a significant role in a rapidly changing 150-year warming trend."[10] Overall, it concludes that extensive efforts to find a natural explanation for current warming have failed, leaving human activity as the primary cause.

Creation of Greenhouse Gases

The main human activity that is causing climate change is the creation of greenhouse gases—primarily carbon dioxide—which are known to trap heat on Earth. It is a natural part of life for carbon dioxide to be both produced and absorbed by Earth. In the past, a balance has been maintained by nature so that levels do not get too high or too low. However, as the EPA explains, human activity is upsetting the natural balance by creating more carbon dioxide than Earth can absorb. Humans are doing this primarily through the mining and burning of fossil fuels, which are used for heating, transportation, manufacturing, and many other societal needs. The EPA says, "Plants, oceans, and soils release and absorb large quantities of carbon dioxide as a part of the Earth's natural carbon cycle. These natural emissions and absorptions of carbon dioxide on average balance out over time. However, the carbon dioxide from human activities is not part of this natural balance."[11] Because humans are creating more carbon dioxide than nature can absorb, it is accumulating at continually increasing levels in the atmosphere, and this is increasing Earth's temperature.

An analysis of historical carbon dioxide levels reveals that carbon is increasing at an unprecedented rate. In order to estimate levels from thousands of years ago, researchers drill samples from deep underneath ice sheets in Antarctica and Greenland. These ice cores contain information about past carbon dioxide levels. More recent levels, dating back to 1958, come from the Mauna Loa Observatory near the Mauna Loa Volcano in Hawaii, the world's oldest continuing monitoring station. According to the Scripps Institution of Oceanography, which conducts the Mauna Loa measuring, during the past eight hundred thousand years levels have fluctuated between 180 parts per million (ppm) [meaning that out of a million air molecules, 180 are carbon dioxide] and 280 ppm, corresponding with ice ages and warm periods in Earth's history. However, since the Industrial Revolution in the nineteenth century, when the use of fossil fuels started to increase, carbon dioxide levels have risen higher and faster than before. In 2013 researchers at Mauna Loa announced that carbon dioxide levels had reached 400 ppm, the highest since monitoring began. Scripps atmospheric scientist Ralph Keeling says, "These are not small changes." Researchers know that these large changes are impacting the climate, however they are unsure how severe the effects will be. Keeling says, "Two or 3 million years ago was the last time we had concentrations in this range, so we're moving into territory that's almost outside the scope of human existence on the planet at this point."[12]

> **"Ninety-seven percent of climate scientists agree that climate-warming trends over the past century are very likely due to human activities."[14]**
>
> —The National Aeronautics and Space Administration, the US government agency responsible for the space program and for aeronautics and aerospace research.

Skeptics Influenced by Business Interests

A small number of people continue to deny that human activity is responsible for the changing climate. Many of these skeptics may have ties to companies that have a strong interest in discrediting the science of climate change. For example, oil and coal–related businesses make billions of dollars from the fossil fuel industry. To continue earning that money they need the public to continue using large amounts of fossil fuels rather than heed the warnings of scientists and reducing fossil fuel use. Before the IPCC's 2013 report on climate change was released, top

UN climate official Halldór Thorgeirsson warned that for this reason, scientists should be ready to see skeptics trying to discredit the report. He says, "Vested interests are paying for the discrediting of scientists all the time. We need to be ready for that."[13]

The evidence clearly shows that human activity is having a significant impact on climate. According to NASA, "Ninety-seven percent of climate scientists agree that climate-warming trends over the past century are very likely due to human activities."[14] And the number one human activity that is contributing to climate change is the creation of greenhouse gases through the use of fossil fuels.

Climate Change Is Not Caused by Human Activity

"[People] somehow came to believe that changes in human activity are causing changes in climate. This is demonstrably false."

—Tomes is a scientist and mathematician.

Ray Tomes, "Human Activity Is Not the Cause of Climate Change, It Is the Result," Cycles Research Institute's Blog, September 17, 2012. http://cyclesresearchinstitute.wordpress.com.

Consider these questions as you read:

1. Do you think that natural variation explains the current period of climate change? Why or why not?
2. How persuasive is the argument that solar variation is responsible for climate change? Explain.
3. How persuasive is the argument that climate change is not caused by human activity? Which facts and ideas are strongest, and why?

Editor's note: The discussion that follows presents common arguments made in support of this perspective, reinforced by facts, quotes, and examples taken from various sources.

Africa's Sahara Desert is the world's largest hot desert. It is one of the harshest environments in the world, with little water or vegetation to support any kind of life. The driest part of this desert is covered with little more than sand and rocks and receives less than an inch of rain per year on average. Sometimes parts of the Sahara go years without any rain at all. However, the Sahara has not always been a desert. Researchers have found evidence that thousands of years ago it had a very different climate. Geologist E. Kirsten Peters says, "What is now the driest part of the Sahara Desert was only four thousand years ago a lush and verdant landscape with lakes, fish, crocodiles, turtles, and people." But the climate did not stay that way. As a result of natural climate variations,

Climate Variation Is a Normal Cycle of the Earth

The current period of warming is not a result of human activity. It is simply natural climate variation, similar to what has occurred all through the earth's history. This graph shows historical temperature information obtained from studying ice cores in Greenland. It reveals that there have been numerous periods of warming. Among these is a period scientists call the Roman Climate Optimum, which occurred during the Roman Empire, and the period they call the Medieval Warming Period, which took place when the Vikings lived in Greenland.

Temperatures of the Last 10,000 Years

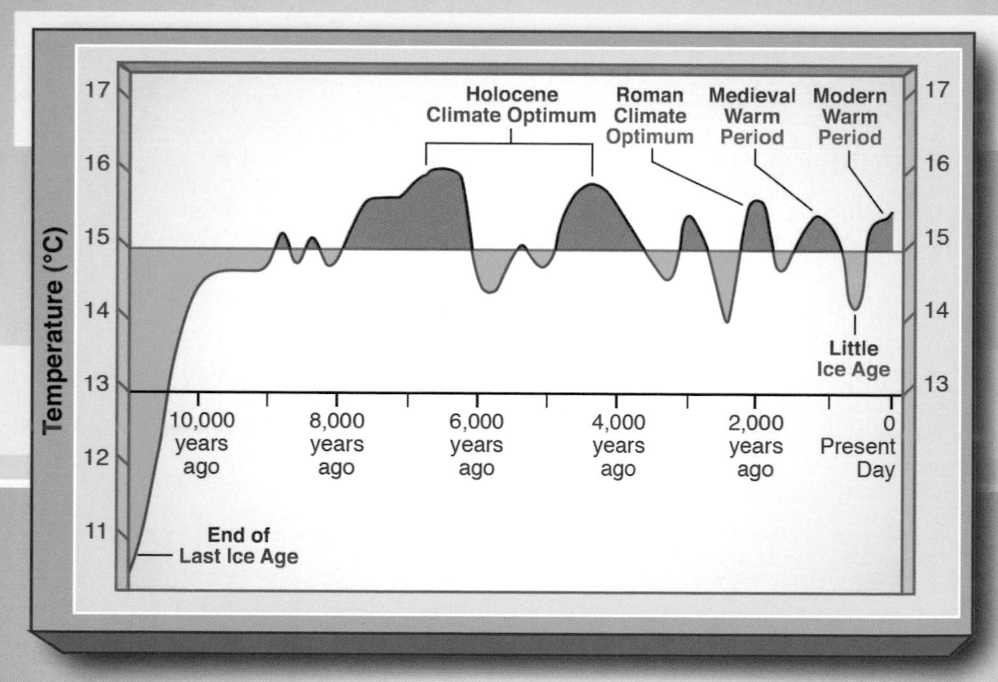

Source: Steve Goreham, "Hot Weather and Climate Change—A Mountain from a Molehill?," *Somewhat Reasonable: The Policy & Commentary Blog of the Heartland Institute*, July 4, 2013. http://blog.heartland.org.

the Sahara was dramatically transformed. Peters explains, "When climate turned yet another corner in Earth's long history, the rains shifted far to the south and the green splendor vanished."[15]

The changing landscape of the Sahara illustrates the way that Earth's climate is never static. Instead, it is continually changing as a result of various natural forces such as fluctuating solar energy and the shifting orbit of

Earth. At present, evidence shows that the climate is changing again. Some people believe that this time the change is due to human activity; however, there is no evidence that this is true. Instead, it is far more likely that current changes are simply a result of natural forces that have changed the climate many times before and will continue to do so in the future.

Cyclic Temperature Variations Are Normal

Analysis of the geologic record shows that major climactic changes are an ongoing process on Earth. No matter what humans do, the climate will continue to go through these cycles of change. As Peters says, "The fact is, if human beings had remained hunter-gatherers throughout our entire history, never producing a single molecule of greenhouses gases through agriculture or industry, climate today would still be changing. It would be lurching toward higher temperatures, crashing toward vastly colder temperatures, or at least swinging toward something different from what has been. That's just the nature of Earth's climate."[16]

> **"The fact is, if human beings had remained hunter-gatherers throughout our entire history, never producing a single molecule of greenhouses gases through agriculture of industry, climate today would still be changing."[16]**
>
> —E. Kirsten Peters, geologist.

Climate change scientists argue that the present day differs from any time in the past because humans are producing such large emissions of greenhouse gases. They insist that these gases must be influencing the climate. However, there is no proof that this is true. Joseph Bast and James M. Taylor of the Heartland Institute argue that at times in the past—long before humans produced substantial amounts of greenhouse gas—the climate was even warmer than it is now. This demonstrates that substantial warming can be caused by natural forces. Numerous scientists researching this topic have found evidence of this natural phenomenon. As Bast and Taylor state, their review of the research shows that "[t]housands of peer-reviewed articles point to natural sources of climate variability that could explain some or even all of the warming in the second half of the twentieth century."[17]

Solar Variation

One possible natural cause of climate change is variation in the amount of energy that the sun releases. Earth gets most of its energy from the sun, and as a result the sun has a significant impact on Earth's climate. Scientists agree that fluctuations in solar energy have changed the climate in the past. PBS Teachers explains, "The amount of energy that flows from the Sun to Earth is so large that even small fluctuations in the Sun's output can cause significant changes in the Earth's climate."[18] One way to measure changes in the sun's energy is to count sunspots. More sunspots indicate that the sun is giving off more energy. PBS Teachers explains how historical sunspot observations show that fluctuations in the sun's energy can have a significant impact on Earth's climate. For example, it says, "During one 30-year stretch in the 1600s—the coldest period of the Little Ice Age when winter temperatures in Europe were from 1 (33.8°F) to 1.5°C (34.7°F) cooler than average—astronomers observed a total of only 50 sunspots, indicating a very quiet Sun. In contrast, the Sun has been more active in recent decades, displaying 160 sunspots or more in one 11-year cycle."[19]

In a 2013 report researchers for the Nongovernmental International Panel on Climate Change reviewed numerous studies about solar activity and climate and also found strong correlations all around the world between solar energy and climate. They concluded that the role of the sun has been greatly underestimated by many climate researchers. They state, "It is fairly certain the Sun was responsible for creating multi-centennial global cold and warm periods in the past." As a result, they insist that changing solar energy is playing an important role in the current warming trend. They argue: "It is quite plausible that modern fluctuations in solar output are responsible for the majority, if not entirety, of the global warming the planet experienced during the past century or so."[20]

Greenhouse Gas Levels Do Not Correlate with Warming

One of the major arguments for human-caused warming is that human emissions of carbon dioxide and other greenhouse gases are trapping heat and causing the world's temperature to rise. However the evidence does not support this theory. Instead, while measurements show that the amount of carbon dioxide in the atmosphere increases every year—hitting an all-time high of 400 parts per million in 2013—global temperatures have not increased substantially for approximately fifteen years. In 2013 congressional testimony, climate researcher Roy W. Spencer argues that this lack of correlation between carbon dioxide levels and warming means there is no strong evidence that human activity is the main cause of recent warming. If carbon dioxide really is a cause of warming, then temperatures should rise as carbon dioxide levels do. However, Spencer says, "The level of warming in the most recent 15 year period is not significantly different from zero, despite this being the period of greatest greenhouse gas concentration." Spencer says, "It is time for scientists to entertain the possibility that something is wrong with the assumptions built into their climate models."[21]

"The level of warming in the most recent 15 year period is not significantly different from zero, despite this being the period of greatest greenhouse gas concentration."[21]

—Roy W. Spencer, climate researcher.

Scientists have studied ice core samples from Antarctica to get information about past temperatures and carbon dioxide levels, and they have found a similar lack of evidence that high carbon levels lead to higher temperatures. These studies actually show that past periods of warming occurred hundreds of years before carbon levels increased. Climate researcher Marc Morano says, "Temperatures lead CO_2 levels on long, intermediate, and short-term timescales." As a result, he says, "CO_2 cannot be the 'control knob' or 'amplifier' of climate, because the tail does not wag the dog, the cause does not follow the effect, and the globe starts to warm and cool 500–5000 years in advance of CO_2 changes."[22]

While the number of humans living on the globe continues to increase, and the world's climate has recently shown evidence of change, it does not follow that human activity is causing that change. Climate change is an ongoing process for Earth and is primarily caused by natural forces, not human emissions of carbon dioxide.

Chapter Two

Has the Extent of Climate Change Been Overstated?

The Extent of Climate Change Has Been Overstated

- Predictions about severe consequences from climate change are exaggerated.
- There has been no significant warming in the past fifteen years.
- Severe weather events have not increased.
- The Intergovernmental Panel on Climate Change has overstated the threat of climate change.

The Extent of Climate Change Has Not Been Overstated

- The world's glaciers and ice masses are rapidly shrinking.
- Sea levels around the world are rising every year.
- World temperatures are increasing and weather patterns are changing.
- The temporary slowing of climate change has not diminished the threat.

The Extent of Climate Change Has Been Overstated

"The warming trend already has stopped and forecasts of future warming are unreliable; and the benefits of a moderate warming are likely to outweigh the costs. Global warming, in other words, is not a crisis."

—Bast is president of the Heartland Institute, an organization that promotes free-market solutions to social and economic problems.

Joe Bast, "Global Warming: Not a Crisis," *Somewhat Reasonable: The Policy and Commentary Blog of the Heartland Institute,* August 10, 2011. http://blog.heartland.org.

Consider these questions as you read:

1. How strong is the argument that because scientific predictions have been wrong, then climate change is not a serious threat? Explain your answer.
2. Do you agree with the argument that severe weather events have not increased? Why or why not?
3. Which pieces of evidence in this discussion provide the strongest support for the view that the extent of climate change has been overstated? Why do you think they are they the strongest?

Editor's note: The discussion that follows presents common arguments made in support of this perspective, reinforced by facts, quotes, and examples taken from various sources.

In 2007 the Intergovernmental Panel on Climate Change issued an alarming report warning that global warming is causing glaciers in the Himalayan mountains to melt. It states, "Glaciers in the Himalaya are receding faster than in any other part of the world and, if the present rate continues, the likelihood of them disappearing by the year 2035 and perhaps sooner is very high if the Earth keeps warming at the current rate."[23]

The Himalayas hold the largest body of water on the planet other than the polar ice caps. Snow and ice-melt from these mountains supply water for hundreds of millions of people. Thus, the disappearance of the Himalayan glaciers would have dire consequences, and the IPCC prediction caused widespread panic. It was later discovered, however, that IPCC's grim prediction was false. While the Himalayan glaciers are decreasing in size like most other glaciers around the world, the rate of melt was vastly overestimated. Scientists now admit that the extent of glacial melt in the Himalayas is poorly understood and needs more study. It is widely agreed that there is no imminent danger of Himalayan glaciers disappearing—and certainly not by 2035. While there is some evidence that Earth is experiencing climate change, this scenario illustrates how greatly the extent of that change has been overstated.

Climate Models Are Flawed

The reality of climate change continues to be minimal and far from the disaster that scientists keep predicting. Scientists have been studying the climate for many years; the IPCC issued its first report in 1990—and scientists have used the group's observations to construct climate models that predict the future. These models show a future in which climate change has serious and harmful consequences, such as the disappearance of the Himalayan glaciers. In reality, however, the actual effects of climate change do not match these alarmist predictions. Instead, the effects have been minimal. Myron Ebell, director of the Centre for Energy and Environment at the Competitive Enterprise Institute, says: "The science contradicts the modellers' dire predictions. The divergence between reality and model projections in the last two decades provides strong evidence that global warming, although it may become a problem some decades in the future, is not a crisis and is highly unlikely to become a crisis." In Ebell's opinion, "We should be worried that the alarmist establishment continues using junk science to promote disastrous policies that will [harm society]."[24]

One prediction that has clearly been exaggerated is the rate that worldwide temperatures will increase. In the 1980s and 1990s there were substantial temperature increases, and scientists predicted that climate change would cause this trend to continue. However these predictions have proved to be greatly overstated. In reality, warming has slowed significantly over the past fifteen years. The rate of increase since 1998 has been so small that it is not even statistically significant. Polling analyst Nate Cohn says, "Since 1998, the warmest year of the twentieth century, temperatures have not kept up with computer models that seemed to project steady warming; they're perilously close to falling beneath even the lowest projections." Society would do well to show a little more skepticism about predictions of huge temperature increases. As Cohn notes, "If scientific models can't project the last 15 years, what does that mean for their projections of the next 100?"[25]

> **"The divergence between reality and model projections in the last two decades provides strong evidence that global warming, although it may become a problem some decades in the future, is not a crisis and is highly unlikely to become a crisis."[24]**
>
> —Myron Ebell, director of the Centre for Energy and Environment at the Competitive Enterprise Institute.

Severe Weather Events

Some researchers have also argued that in addition to increasing temperatures, severe weather events are getting more common and will continue to do so as a result of climate change. Once again, this is an example of exaggeration. Roy W. Spencer says, "There is little or no observational evidence that severe weather of any type has worsened over the last 30, 50, or 100 years." Severe weather events have had too much variation to form any sort of distinct pattern that can be tied to a single cause. As Spencer explains, "Long-term measurements of droughts, floods, strong tornadoes, hurricanes, severe thunderstorms etc. all show no obvious trends, but do show large variability from one decade to the next, or even one year to the next. While the 2003 heat wave in France and the 2010 heat wave in Russia were exceptional, so were the heat waves of the 1930s in the U.S., which cannot be blamed on our greenhouse gas emissions."[26]

Climate researcher Roger Pielke Jr. agrees. In 2013 congressional testimony at a Senate hearing on climate change he reports on severe weather events in the United States and finds that hurricanes have not increased in frequency or intensity since at least 1990 and floods and tornadoes since at least 1950. In fact, he says, flood losses as a percentage of the United States' gross domestic product have actually dropped by approximately 75 percent since 1940.

Public views on this subject may be skewed by events of the moment and by lack of perspective on the past. Severe weather events might *seem* more common now, says climatologist and chair of the School of Earth and Atmospheric Sciences Judith A. Curry, but only because people forget the many severe events that have occurred previously. She says, "In the U.S., most types of weather extremes were worse in the 1930's and even in the 1950's than in the current climate." But then weather patterns calmed down in the 1970s, which is why there seems to be such a contrast between storms of the recent past and today. As Curry says, "The weather was overall more benign in the 1970's. This sense that extreme weather events are now more frequent and intense is symptomatic of 'weather amnesia' prior to 1970."[27]

"In the U.S., most types of weather extremes were worse in the 1930's and even in the 1950's than in the current climate."[27]

—Judith A. Curry, climatologist and chair of the School of Earth and Atmospheric Sciences in Georgia.

Overstatement by the IPCC

Some of the blame for exaggeration about the extent of climate change lies with the IPCC. The IPCC is widely believed to be an authority on climate change, and its predictions have a significant impact on public opinion about the topic. However there is evidence that the organization has exaggerated claims about the extent of the problem. For example, Paul C. "Chip" Knappenberger of the Center for the Study of Science at the Cato Institute says the IPCC is wrong in its predictions about how greenhouse gases will cause temperatures to rise. According to Knappenberger, "The latest climate science (some 10 studies published in just the

No Evidence for Increasing Heat Extremes

While many scientists warn that climate change is causing increases in extreme weather events such as heat waves, this graph of US temperatures does not support that theory. The graph shows the number of daily record-high temperatures for summertime since 1895. There is no trend of an increasing number of record temperatures. The highest numbers of record temperatures actually occurred in the 1930s.

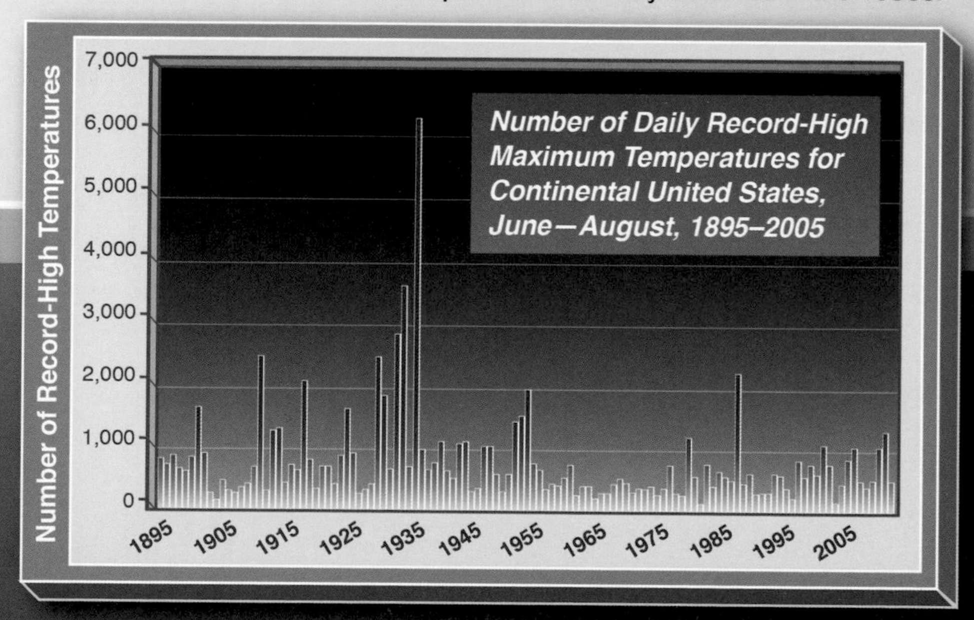

Source: Judith A. Curry, "Statement to the Committee on Environment and Public Works of the United States Senate," hearing on "Review of the President's Climate Action Plan," January 16, 2014. http://scienceandpublicpolicy.org.

past 3 years) indicates that the earth's climate sensitivity—that is, how much the global average surface temperature will rise as a result of greenhouse gases emitted from human activities—is some 30 percent *less* than scientists thought at the time of the last IPCC Assessment, published in 2007." In addition to being wrong about the impact of greenhouse gases, he argues that the IPCC is also overstating projections of future climate change because the climate change models it uses to predict the future are incorrect. He says, "The IPCC's projections of future climate change and the resulting impacts are nearly twice as large as they should be. In other words, the models don't work."[28]

The modelers—the people who interpret the climate models—are suspect as well. Many of the people who create the IPCC reports are environmentalists who believe that society needs to change its behavior because it is destroying the environment. These environmentalists have a tendency to interpret climate data to support their own beliefs. Steven F. Hayward, senior fellow at the Pacific Research Institute and author of the *Almanac of Environmental Trends,* says, "The people who participate in the IPCC process are very knowledgeable people, but they tend to be self-selected as believers in the catastrophic global warming hypothesis. I know from personal experience that a number of them are simply environmental crusaders."[29]

Overall, the extent of climate change has been overstated. While the IPCC and other researchers have made dire predictions about how the climate is changing, in reality there is no evidence for such claims.

The Extent of Climate Change Has Not Been Overstated

"The Earth is warming. . . . Such warming carries with it a risk of serious, perhaps even catastrophic impacts for humans and the rest of the planet's ecosystems."

—Dessler is a climate scientist and a former senior policy analyst in the White House Office of Science and Technology Policy.

Andrew E. Dessler, *Introduction to Modern Climate Change*. New York: Cambridge University Press, 2012, p. 165.

Consider these questions as you read:

1. Do you think society will be harmed by declines in the world's ice? Why or why not?
2. How strong is the argument that temperature increases pose a serious threat? Explain your answer.
3. Do you agree that the action of the ocean to absorb heat is only slowing climate change temporarily? Why or why not?

Editor's note: The discussion that follows presents common arguments made in support of this perspective, reinforced by facts, quotes, and examples taken from various sources.

While very few people live in Antarctica, this continent has a huge impact on life all over the world. *National Geographic* explains, "In many ways, Antarctica is the planet's great thermostat, driving climate with its cycles of freeze and thaw."[30] Every winter a huge ring of ice forms around Antarctica, and every summer that ice melts. According to *National Geographic*, "This melt and release affects ocean current circulation, redistributes the heat of the sun, and regulates climate, affecting the planet's weather—and our lives—at the most fundamental level."[31]

However, climate change threatens to dramatically change this cycle. Research shows that overall, Antarctic ice is significantly decreasing. In 2002 the Larsen B Ice Shelf—an ice shelf on the east coast of Antarctica approximately the size of the state of Rhode Island—disintegrated. Geology expert Donald R. Prothero explains that such a collapse is unprecedented. He says, "The Larsen B shelf had survived all the previous ice ages and interglacial warming episodes for the past 3 million years, and even the warmest periods of the last 10,000 years—yet it and nearly all the other thick ice sheets on the Arctic, Greenland, and Antarctic are vanishing at a rate never before seen in geologic history."[32] Vanishing Antarctic ice will have a significant impact on the continent's natural melting cycle and in turn will impact the climate of the entire globe. This is only one example which shows that the extent of climate change has not been overstated. In reality, climate change is having a significant impact on the planet and poses a serious threat to humankind.

Vanishing Ice and Rising Sea Levels

Antarctica is not the only part of the world that is losing ice. Glaciers almost everywhere are retreating, and two of the world's other large ice masses, the Greenland ice sheet and Arctic sea ice, have also declined. For hundreds of years the Arctic Ocean has been almost completely covered by ice in the winter. Some of that ice melts in the summer, but it always refreezes in winter. Now climate change is resulting in less ice every winter. Writer Michael Le Page explains how the situation is getting progressively worse. He says, "As the world has warmed in the past decades, the winter refreeze has stopped compensating for the summer melt. Heat-reflecting white ice has given way to heat-absorbing dark water; snow has melted even earlier on surrounding lands; more heat-trapping moisture has entered the atmosphere; and bigger waves and storms have assailed weakening ice." According to Le Page, Arctic sea ice had reached its lowest level in at least fourteen hundred years by the end of the 1990s. He says, "At the end of this summer [2012] only a quarter of the Arctic Ocean was still covered in ice, a record low in modern times, and the total volume of ice was just a fifth of what it was three decades ago."[33]

Researchers fear that the world is headed toward a time when there will be no more ice in the Arctic. If this comes to pass, it will spell disaster for the human and animal inhabitants of the region. The decline of Arctic sea ice also poses a threat to the rest of the world because as that ice melts, worldwide sea levels are rising. Sea levels are already rising in countries around the world. The American Meteorological Society reports that the average global sea level has risen by approximately seven inches in the twentieth century and that the rate of increase has been accelerating since the early 1990s. Scientists fear that if the world's ice continues to melt at the rapid pace it has been following, then sea-level rises will be very large. Radley Horton, a research scientist at Columbia University's Earth Institute in New York City, says that in the past few years, the ice sheets in Greenland and West Antarctica have been melting at an accelerated pace. He says, "The concern is that if the acceleration continues, by the time we get to the end of the 21st century, we could see sea-level rise of as much as six feet globally."[34]

"The concern is that if the acceleration continues, by the time we get to the end of the 21st century, we could see sea-level rise of as much as six feet globally."[34]

—Radley Horton, research scientist at Columbia University's Earth Institute in New York City.

Increasing Temperature

Climate change is also steadily increasing Earth's temperatures. According to the IPCC, temperatures have been rising over the past thirty years. It says, "Each of the last three decades has been successively warmer at the Earth's surface than any preceding decade since 1850." While Earth's weather continues to fluctuate from month to month, including periods of both heat and cold, the overall trend is toward a warming planet. The IPCC says, "In the Northern Hemisphere, 1983–2012 was *likely* the warmest 30-year period of the last 1400 years."[35]

The higher temperatures can also help to explain changing US snow patterns. In the nation's western mountains, rain has replaced snow at elevations where snow was once commonplace. According to the American

Warming Is Causing Arctic Ice to Shrink

Climate change is causing the world's large ice fields to shrink every year. Since these ice fields affect world sea levels and weather patterns, this decline will have significant implications for the entire world. The graph shows the extent of sea ice in the Arctic during the summer, from 1900 to the present, revealing that it is steadily declining. Overall, sea ice is at a far lower level than it was in 1900.

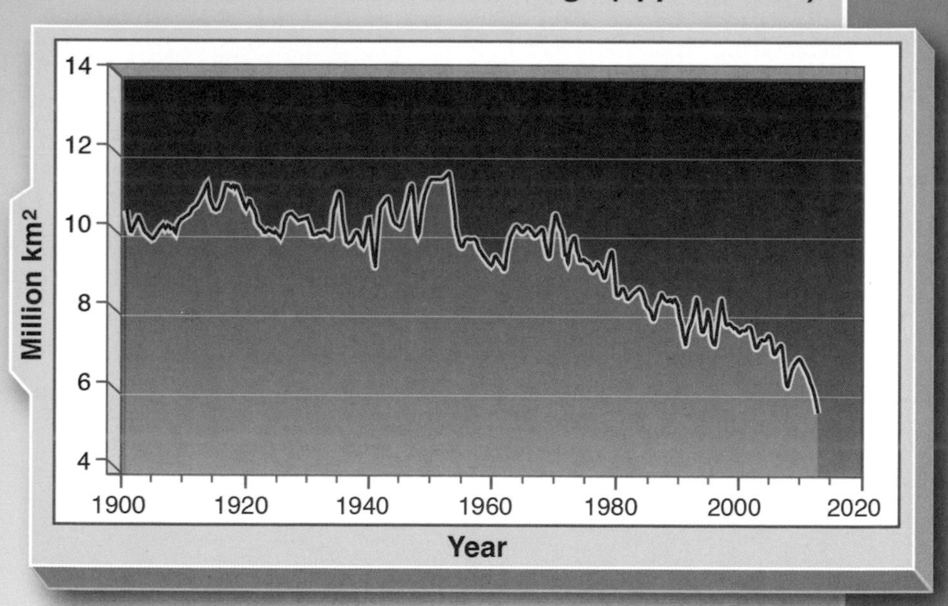

Source: Intergovernmental Panel on Climate Change, "Climate Change 2013: The Physical Science Basis; Summary for Policymakers," 2013. www.climatechange2013.org.

Meteorological Society, snowpacks in the mountains have decreased and are melting earlier. This situation means less water for people and wildlife in a large swath of the western United States. But the effects of higher temperatures are not just being felt in the west. Many parts of the country are experiencing longer frost-free periods, earlier springs, longer growing seasons, and changes in the natural habits and migratory patterns of insects and birds, the meteorological society notes.

Pause in Warming Is Temporary

Some critics point out that the pace of warming has recently slowed, and as a result they argue that the extent of climate change has been overstated. However, this is incorrect. The pace of warming has only slowed temporarily, due to the ocean absorbing a large amount of heat. Climate change is still a serious problem. It is causing substantial changes to Earth, and the ocean is only temporarily slowing some of these changes. Yair Rosenthal and other researchers at Rutgers University published a study in *Science* in 2013 explaining the role of the ocean in climate change. They estimate that the middle depths of the Pacific Ocean had warmed approximately fifteen times faster in the past sixty years than at any other time in the past ten thousand years. Rosenthal says this is a likely reason why climate change has recently slowed below what was predicted. He says, "We may have underestimated the efficiency of the oceans as a storehouse for heat and energy." Overall, however, he warns that climate change is still a serious problem. He says, "It may buy us some time—how much time, I really don't know. But it's not going to stop climate change."[36]

Volcanic activity has also caused warming to slow temporarily. Since 2005, volcanic activity has put a large amount of sulfur particles into the air. These particles have a cooling effect because they reflect some of the sun's energy back into space. Susan Solomon, an atmospheric scientist at the Massachusetts Institute of Technology, believes that this volcanic activity is a significant factor in the recent slowing of warming. She says "The measurements are very clear that it's part of the story—maybe 25 percent, maybe 30 percent."[37] However, like the impact of the ocean, volcanic activity is causing only a temporary pause in warming and has not diminished the seriousness of that warming.

> **"Our planet is changing in ways that will have profound impacts on all of humankind."[38]**
>
> —Barack Obama, forty-fourth president of the United States.

The extent of climate change has not been overstated. As US president Barack Obama says, "Our planet is changing in ways that will have profound impacts on all of humankind."[38] These changes include rapid declines in ice, rising sea level, and substantial changes in weather.

Chapter Three

Will Climate Change Be Harmful to Society?

Climate Change Will Be Harmful to Society

- Climate change will alter the weather and cause an increase in extreme weather events.
- The sea level will rise and threaten communities near the coast.
- Changing ecosystems will displace plants, animals, and people.
- A changing climate will increase the spread of disease.

Climate Change Will Not Be Harmful to Society

- Society has proven that it can adapt to significant changes in climate.
- Plants and animals will successfully adapt as the climate changes.
- Increased levels of carbon dioxide will actually increase worldwide agricultural yields.
- There is no evidence that climate change will cause extreme weather to increase.

Climate Change Will Be Harmful to Society

"Climate change is altering people's lives, right now, from the United States to Africa to the Arctic. It is as clear and present a danger as we've ever seen. And the threat will only get worse if left unchecked."

—Hansen runs the Climate Science, Awareness and Solutions program at the Columbia University Earth Institute. He is former director of the NASA Goddard Institute for Space Studies.

James Hansen, "What We Owe Our Kids on Climate," CNN, December 6, 2013. www.cnn.com.

Consider these questions as you read:

1. Do you agree with the argument that extreme weather events threaten society? Why or why not?
2. Do you agree that climate change will displace plants and animals? Why or why not?
3. How persuasive is the argument that climate change will be harmful to society? Which piece of support is the weakest, and which is the strongest?

Editor's note: The discussion that follows presents common arguments made in support of this perspective, reinforced by facts, quotes, and examples taken from various sources.

More than thirty-two hundred years ago the eastern Mediterranean coast was populated by numerous flourishing civilizations including the Hittites, the Aegeans, and the Egyptians. These groups of people had sophisticated cultures with busy systems of trade and large cities. They made important advances in science, medicine, and technology. Then within a short period of time, many of these powerful civilizations abruptly collapsed and disappeared from history. What happened has long been a mystery. However, in 2013 researchers revealed the results of a new study

Climate Change Will Cause a Harmful Sea-Level Rise

Climate change is causing the world's sea level to gradually rise, a phenomenon that could eventually lead to massive flooding and the displacement of millions of people in the United States alone. This chart shows the potential impacts on the American population, based on different scenarios calculated by scientists. According to these projections, the homes of at least 1.3 million people and as many as 7.8 million could experience flooding by rising seas in the future.

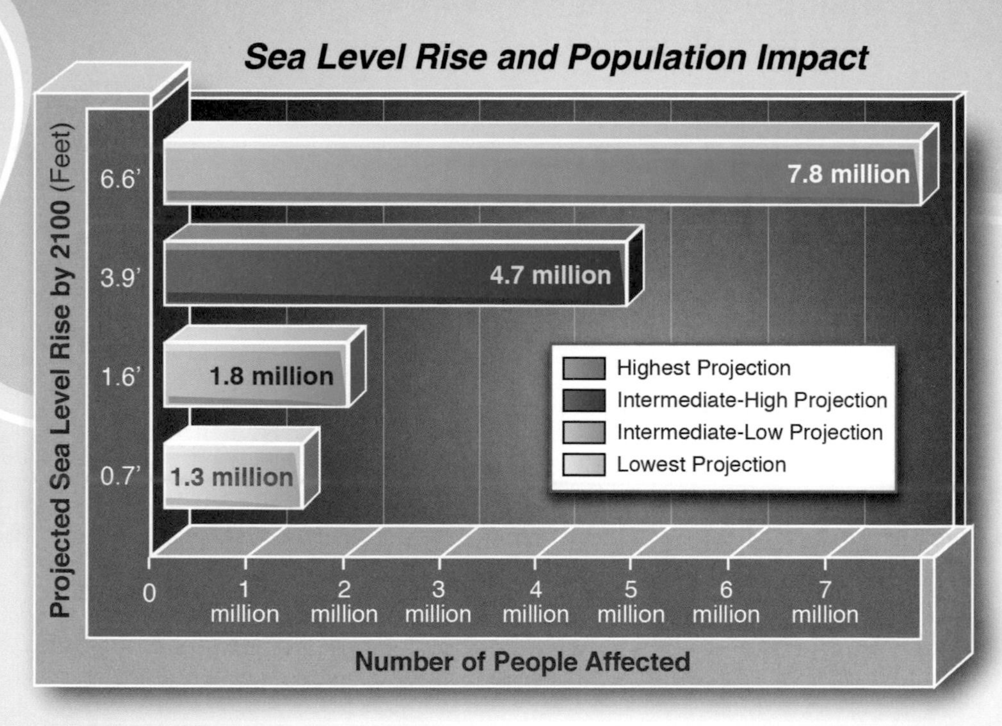

Source: Heidi Cullen, "Extreme Weather Within the Context of Our Changing Climate," Briefing to the United States Senate Committee on Environment and Public Works, July 18, 2013.

which suggest that climate change was the likely reason for the collapse. By analyzing grains of pollen from the area, they concluded that a changing climate caused a huge drought in the area, disrupting farming and trade and making it impossible for these large societies to continue. As this example shows, climate change poses grave threats to human civilization. Just as a changing climate contributed to the downfall of past civilizations, the present period of climate change threatens to severely disrupt the way of life for many people around the world.

Extreme Weather Events

Climate change will harm society in many ways. One is by causing extreme weather events such as heat waves or storms. Climate change makes these events more likely because it disrupts the forces that determine the world's weather, resulting in unpredictable and extreme events. There is evidence that these extreme events are becoming increasingly common. For example, in 2010 Russia experienced an extended summer heat wave that broke all temperature records. Thousands of people died from heat stress and respiratory illnesses. Andrew E. Dessler, a climate scientist and former senior policy analyst in the White House Office of Science and Technology Policy, says, "The Russians learned the hard way that warmer temperatures do not mean tank tops and grilled hot dogs, but instead mean wildfires, loss of agricultural crops, and human suffering."[39]

Experts believe that devastating events such as this will become more common as the climate warms. For example, in the case of Russia, according to Reuters news reporter Deborah Zabarenko, "Computer models show the risk of such heat waves in western Russia could rise from less than 1 percent in 2010 to 10 percent or more by 2100 as the concentration of greenhouse gases in the atmosphere increases."[40]

"A record 14 weather disasters occurred in 2011, sustaining more than $1 billion each in economic losses for a total of $60.6 billion."[41]

—Coral Davenport, journalist.

Not only do extreme weather events cause a loss of human life but they cost society billions of dollars in lost revenues and damages. Economic losses due to the Russian heat wave were estimated at more than $15 billion. In the United States, extreme weather events such as tornadoes, droughts, and floods have also been extremely costly. According to journalist Coral Davenport, "In the United States, 2011 and 2012 were the two most extreme years on record for destructive weather events." She says, "A record 14 weather disasters occurred in 2011, sustaining more than $1 billion each in economic losses for a total of $60.6 billion."[41]

Rising Sea Level

The rising sea level that is resulting from climate change will also be harmful to society. Hundreds of millions of people around the world live in low-lying islands or coastal areas. For example, in 2010, according to NOAA's most recent statistics, 39 percent of the US population lived in counties directly on the shoreline. The rising sea will destroy these coastal communities and important ecosystems, such as estuaries. It could also contaminate groundwater supplies with saltwater and threaten landfill and hazardous-waste sites located at the coast.

While destruction from rising sea levels is a future threat for many coastal communities, in some places that destruction is already occurring. The Republic of Kiribati in the Pacific Ocean comprises thirty-three islands, all but one of which are low-lying atolls. The World Bank says it is one of the most vulnerable countries in the world to climate change and sea-level rise. As the sea level has risen in recent years, it has already destroyed villages and farmland in Kiribati, and saltwater is seeping into the islands' underground freshwater supplies. According to *Bloomberg Business Week* columnist Jeffrey Goldberg, "If scientists are correct, the ocean will swallow most of Kiribati before the end of the century, and perhaps much sooner than that."[42] Kiribati is also one of the poorest nations in the world; it does not have the resources to build seawalls or take other steps to fight the rising sea. As a result, the country's 103,000 citizens are at risk of losing their entire country.

Impact on Wildlife

Displacement resulting from climate change would affect more than people. The changing climate will also impact plants and animals that depend upon specific conditions to survive. As conditions change, these plants and animals will be forced to relocate, adapt to changing conditions, or face extinction. In research published in 2013 in the journal *Nature Climate Change*, Rachel Warren and a group of researchers from the University of East Anglia in the United Kingdom studied fifty thousand globally widespread and common species. They found that more than one-half of the plants and one-third of the animals will lose more than half of their climatic range—meaning the areas they can survive in—by 2080 if nothing is done to slow climate change. For example, warming temperatures will push some animals to higher elevations and others toward coastlines, eventually leaving them nowhere else to go.

The movement or extinction of plants and animals will have a significant impact on entire ecosystems. Ecosystems depend on a complex balance, and even small changes can have substantial effects. For example, the EPA explains how ice algae in the Arctic are declining as a result of climate change, and this is having a far-reaching effect. It says, "These algae are eaten by zooplankton, which are in turn eaten by Arctic cod, an important food source for many marine mammals, including seals. Seals are eaten by polar bears. Hence, declines in ice algae can contribute to declines in polar bear populations."[43]

"When global temperatures rise . . . the areas where malaria is endemic will expand into regions where malaria has never been considered a health concern."[45]

—Bradley J. Dibble, physician.

Organisms in ecosystems are linked together in extremely complex webs and are thus interdependent on one another. Because humans are also part of the ecosystem, changes to that system will threaten them too. Writer Michael Hutchins uses an analogy to explain the impact of species loss, stating, "Well-known biologists Paul and Anne Ehrlich once likened this to taking the bolts out of a flying aircraft one at a time. It may hold together for a while, but eventually, a wing will fall off and the entire plane will crash."[44]

Increases in Disease

Yet another way that climate change threatens society is by impacting human health. The World Health Organization argues that health impacts from climate change will be overwhelmingly negative. One impact will be to increase the spread of vector-borne diseases such as malaria and dengue fever, which already pose a significant health problem around the world. These diseases are spread by blood-sucking arthropods (insects or arachnids). For example, mosquitos spread malaria to humans and ticks spread Lyme disease. Scientists believe that as temperatures increase, the range of where these vectors can live will also increase, meaning that more people will be vulnerable to the diseases they spread. Physician Bradley J. Dibble talks about the example of mosquitos, already a major global health problem. Mosquitos live in tropical regions of the world. He says, "When global temperatures rise, the area where the mosquito vector can thrive will expand. . . . The areas where malaria is endemic will expand into regions where malaria has never been considered a health concern, with more cases added to the list and more deaths as a result." In addition, he adds, "Hotter weather will lead to longer summers and, therefore, longer mosquito seasons."[45]

Overall, climate change will be extremely harmful to society. It will change the environment through effects such as altered weather and rising seas, and it will drastically impact ecosystems and human health.

Climate Change Will Not Be Harmful to Society

"Any warming that may occur is likely to be modest and cause no net harm to the global environment or to human well-being."

—The Nongovernmental International Panel on Climate Change is an international panel of scientists and scholars that studies climate change.

The Nongovernmental International Panel on Climate Change, "Climate Change Reconsidered II: Physical Science: Executive Summary," 2013. http://climatechangereconsidered.org.

Consider these questions as you read:

1. Do you agree that humans are extremely adaptable to climate changes? Why or why not?
2. How strong is the argument that increased carbon dioxide will be benefit agriculture? Explain your opinion.
3. Which pieces of evidence in this discussion provide the strongest support for the argument that climate change will actually be beneficial to society? Why do you think they are the strongest?

Editor's note: The discussion that follows presents common arguments made in support of this perspective, reinforced by facts, quotes, and examples taken from various sources.

Before the world even started to worry about the threat of rising seas from climate change, the Netherlands was taking action to protect itself from the sea. The Netherlands is an extremely low-lying country; approximately 26 percent of it is actually below sea level. This means that any changes in the level of the sea have the potential to be devastating. In 1953 the nation realized this after a storm caused the ocean to flood the southwest part of the country, killing 1,835 people in a single night. In addition, an estimated thirty thousand animals drowned and more than forty-seven thousand buildings were damaged. The flood cost the Netherlands millions of dollars, but since then the country has actively fought back against the sea. At present a network of dams, sluices, and barriers successfully protects its citizens even though many of them live below sea level. The Netherlands has even reclaimed land from the ocean by building dikes and pumping out sea water. The Dutch example shows that climate change does not have to be harmful to society. Humans have the ability to successfully adapt to a changing environment, just as the Dutch have done.

Adaptation

Humans are a highly adaptable species. For example, some individuals make their homes in Death Valley, which often reaches 120°F (49°C) in the summer. Others call the Arctic Circle home despite winter temperatures around -30°F (-34°C). If Earth's climate undergoes huge changes in the future, this ability to adapt to a wide variety of conditions will allow humans to survive. Technology is likely to play a part in this adaptation. Due to technological development, over time society has already become less sensitive to climate changes. For example, technology allows US society to predict and prepare for extreme weather events, and deaths due to extreme weather events have decreased as a result. The Cato Institute argues that rather than being negatively affected by climate change, humans will actually continue to become less affected by it in the future. It predicts, "If future populations are wealthier than current ones and technological advances continue, as has been the case since the Industrial Revolution, they should be more resilient and less vulnerable to the adverse effects of climate change."[46] Journalist Anne Jolis gives an illustration of how technological advances have made society far less sensitive to extreme weather. In 2011 a rare storm affected parts of the United States, including Texas, which was at the time preparing to host the Super Bowl in Dallas. Jolis says, "[The] storm wreaked havoc and left hundreds of football fans stranded, cold, and angry. But thanks to modern infrastructure, 21st century health care, and stockpiles of magnesium chloride [used to melt snow] and snow plows, the storm caused

Increased Carbon Dioxide Levels Will Benefit Agriculture

Rising atmospheric carbon dioxide levels will lead to greater agricultural yields—a clear benefit to society. This graph shows the percentage by which plant growth is enhanced when atmospheric carbon dioxide levels are enriched to various levels above normal. It is based on data from 1,087 individual experiments, representing a wide range of species. Even when carbon dioxide increases to very high levels, the graph shows higher yields.

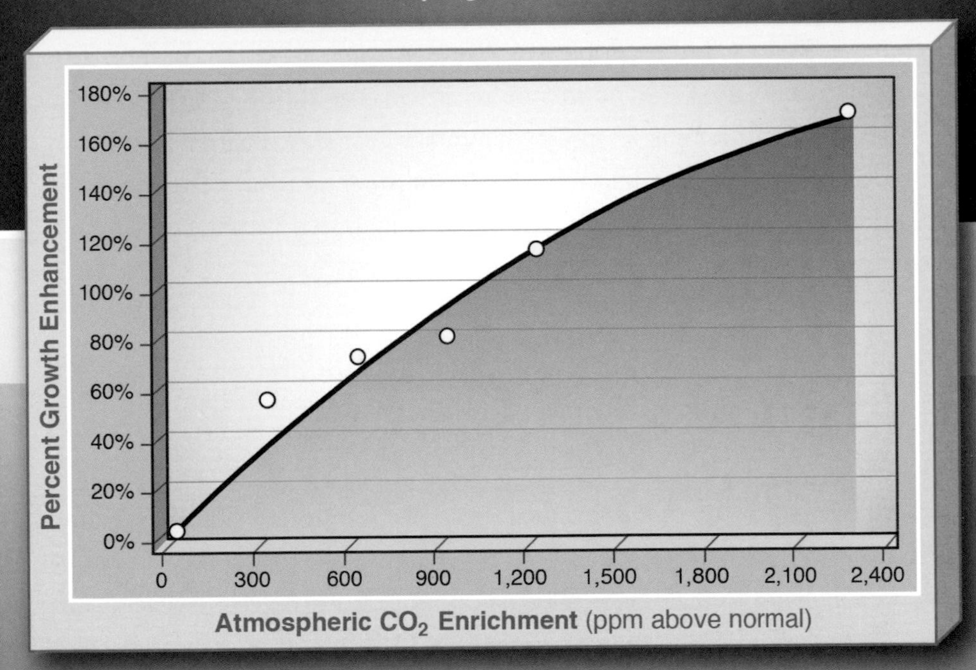

Source: Cato Institute, Robert C. Balling et al. "Addendum: Global Climate Change in the United States," October 31, 2012. www.cato.org.

no reported deaths and Dallas managed to host the big game on Sunday." Jolis says, "Compare that outcome to the 55 people who reportedly died of pneumonia, respiratory problems and other cold-related illnesses in Bangladesh and Nepal when temperatures dropped to just above freezing last winter [2010]."[47] These people did not have the technology that the United States does to protect them from weather changes.

In fact, warming trends will actually benefit some groups of people, such as those who live in cold parts of the world, where the climate makes it more difficult to survive. Research shows that cold weather poses a far greater threat to people's health than warm weather does. For example, in a 2007 study University of California researchers Olivier Deschenes and Enrico Moretti found that extreme cold kills more people in the United States than extreme heat; in fact it is the cause of more deaths than homicide. The researchers report that in recent years there has been a trend of migration from cold Northeastern states to warm Southwestern states. They estimate that this migration has had a huge effect on longevity, delaying the deaths of approximately 5,400 people every year. Warmer weather can have economic benefits too. The authors of an article that appeared in the online German magazine *Spiegel* argue, "Countries like Canada and Russia can look forward to better harvests and blossoming tourism. The countries bordering the Arctic also hope that the melting of sea ice will enable them to reach previously inaccessible natural resources."[48]

Plants and Animals

Like people, plants and animals also have a remarkable ability to adapt to climate changes. This can be seen in coral reefs. Scientists have observed that increased levels of carbon dioxide are leading to rises in temperature and acidity in the world's oceans. These changes have prompted fears of irreversible harm to coral reefs. But the opposite has occurred in places such as Kaneohe Bay on the Hawaiian island of Oahu. Journalist James Vlahos says, "Because of unusual water circulation, the bay is more acidic and a few degrees warmer than the regional norm. This is a climate-change–like scenario that most Hawaiian coral won't confront until midway through the 21st century or later—and yet, as the Hawaii Institute of Marine Biology's Christopher Jury reported at the ICRS [International Coral Reef Symposium] conference, 'growth is strong and reef development robust.'"[49] Jury is among researchers who believe that coral might be able to adapt to changes in temperature and acidity as the climate warms.

Plants also have the ability to adapt to changing conditions. Further, the increased carbon dioxide levels that are expected in the future are actually likely to be beneficial to plants. Many scientists insist that plants grow better with higher levels of carbon dioxide, and these scientists have conducted numerous experiments to prove it. In a 2013 *Wall Street Journal* editorial one group of scientists states, "[Carbon dioxide is] a key component of the biosphere's life cycle. Plants do so much better with more CO_2 that greenhouse operators often increase the CO_2 concentrations by factors of three or four to get better growth."[50] They point out that worldwide agricultural yields have increased substantially in the past century. While some of that is due to improvements in farming, it is also due to the fact that carbon dioxide levels have increased. Scientists believe that if carbon dioxide levels continue to increase in the future, agriculture will become increasingly productive and food shortages less of a problem.

> **"Numerous studies conducted on hundreds of different plant species testify to the very real and measurable growth-enhancing, water-saving, and stress-alleviating advantages that elevated atmospheric CO_2 concentrations bestow upon Earth's plants."[51]**
>
> —Craig D. Idso, founder of the Center for the Study of Carbon Dioxide and Global Change.

Likewise, the Center for the Study of Carbon Dioxide and Global Change examines the effect of rising carbon dioxide levels on global food production and finds that rising CO_2 is extremely beneficial for plants and is increasing food production. It says, "Numerous studies conducted on hundreds of different plant species testify to the very real and measurable growth-enhancing, water-saving, and stress-alleviating advantages that elevated atmospheric CO_2 concentrations bestow upon Earth's plants."[51] The organization predicts that society will actually see monetary benefits in the future due to increased agricultural yields as carbon dioxide levels increase.

No Evidence That Extreme Weather Is Increasing

Finally, worries that climate change is causing extreme weather events such as droughts, heat waves, and storms are largely based on speculation rather than on hard evidence. Researchers simply do not have enough accurate information about the history of extreme weather events to accurately assess whether they are becoming more severe. The Nongovernmental International Panel on Climate Change argues, "Although specific regions have experienced significant changes in the intensity or number of extreme events over the twentieth century, for the globe as a whole no relationship exists between such events and global warming over the past 100 years."[52]

"There's no data-driven answer yet to the question of how human activity has affected extreme weather."[53]

—Roger Pielke Jr., climate researcher at the University of Colorado.

Roger Pielke Jr., a University of Colorado climate researcher, agrees that at present researchers simply do not have enough information to make a prediction about future occurrences of extreme weather. He insists, "There's no data-driven answer yet to the question of how human activity has affected extreme weather."[53]

Climate change will not be harmful to society because people, plants, and animals all have the ability to adapt to changing conditions. In some cases, changes may even be beneficial.

Chapter Four

Should Society Try to Reduce Climate Change?

Society Should Try to Reduce Climate Change

- Society should reduce its carbon emissions to help slow climate change.
- US action on climate change is important.
- The most effective way to address climate change is to reduce the use of fossil fuels.
- Geoengineering may be an effective solution to climate change.

Society Should Not Try to Reduce Climate Change

- It is impossible for society to do anything that will reduce climate change.
- Using less fossil fuel will have little impact on the climate and will be harmful to society.
- Geoengineering is too risky and should not be attempted.
- Society should plan for adaptation rather than try to reduce climate change.

Society Should Try to Reduce Climate Change

"If our changing climate goes unchecked, it will have devastating impacts. . . . Responding to climate change is an urgent public health, safety, national security, and environmental imperative."

—McCarthy is the administrator of the US Environmental Protection Agency, the federal government agency that works to protect human health and the environment.

Regina McCarthy, opening statement, "Hearing on the Obama Administration's Climate Change Policies and Activities," Subcommittee on Energy and Power, Committee on Energy and Commerce, US House of Representatives, September 18, 2013. http://democrats.energycommerce.house.gov.

Consider these questions as you read:

1. Do you agree that society needs to substantially reduce its carbon emissions? Why or why not?
2. Can you think of any possible harm that might result from using increased amounts of alternative energy such as wind power? Explain.
3. How strong is the argument that geoengineering might be a good solution to climate change? Explain.

Editor's note: The discussion that follows presents common arguments made in support of this perspective, reinforced by facts, quotes, and examples taken from various sources.

Countries can and must try to halt climate change. They can do this by reducing greenhouse gas emissions. This is exactly what a fifteen-member group of European countries called the EU-15 have done. The EU-15, which includes France, Germany, and the United Kingdom, are signatories of the Kyoto Protocol. Under this international treaty 192 industrialized countries have agreed to legally binding obligations to reduce their greenhouse gas emissions. The agreement went into force in 2005, and the European Commission reports that as of 2011—the latest year for which data is available—the EU-15 countries have succeeded in reducing emissions. Under the Kyoto Protocol, countries agreed to reduce emissions 8 percent below the level of a certain base year, which is 1990 in most cases. In 2011 EU-15 emissions were 14.9 percent below their base level, meaning that these countries greatly surpassed their goal. As this example shows, action against the threat of climate change is possible. Society should follow the example of the European Union, and take action now to reduce climate change.

Reduce Carbon Emissions

The most effective action society can take to address climate change is to reduce emissions of carbon dioxide—the primary greenhouse gas—since this is one of the main causes of warming. In the United States large amounts of carbon come from the combustion of fossil fuels for transportation and also from power plants that provide energy for homes, businesses, and factories. This means that successful emissions reduction must focus on these two areas. US president Barack Obama insists that the nation must impose strict limits on the amount of carbon these power plants can emit. He points out that there are limits on many other kinds of pollutants, and it makes sense to limit carbon dioxide emissions too. He says, "Today, about 40 percent of America's carbon pollution comes from our power plants. But here's the thing: Right now, there are no federal limits to the amount of carbon pollution that those plants can pump into our air. None. Zero." He says, "That's not right, that's not safe, and it needs to stop."[54]

US Action Is Important

In the fight to reduce climate change, the actions of the United States will be an important factor in determining success or failure. This is because the United States is a big part of the problem; in 2011 it produced more carbon dioxide emissions than any other country except China. The United States also has the power to influence the actions of other countries. Richard W. Miller, author of *God, Creation, and Climate Change*, argues that at present, the United States is making the problem of climate change worse because it has refused to participate in international agreements to reduce emissions. In his opinion, "The United States has been the chief obstacle to all the significant attempts to avoid this fate [of climate change]." He says, "Despite being the largest historical emitter, responsible for 27 percent of the fossil fuel CO_2 emissions in the atmosphere since 1750 (China is second at 9 percent and India third at 2.7 percent), the United States has been the main barrier to moving the world away from a fossil-fuel-based economy. We never signed the Kyoto Protocol [to reduce greenhouse gases], and our credibility in international negotiations, such as those at Copenhagen [climate change summit] in December 2009, has been undermined by our failure to pass domestic climate legislation."[55] If the world is to take successful action against climate change, the United States must start setting an example to the world rather than continuing its refusal to act.

However, in order to make a substantial reduction on worldwide carbon emissions, large developing nations such as China and India must also take action to reduce their emissions. While developed countries such as the United States have actually decreased their carbon dioxide emissions in recent years, emissions are rapidly increasing in developing nations because they are using large amounts of fossil fuels to drive their economic growth. For example, according to the International Energy Agency, in 2011 carbon dioxide emissions in China grew by 9.7 percent. India's emissions are also substantial. Adam Sieminski, administrator of the US Energy Information Administration, says that in projections of future energy use, "These two countries combined account for half the world's total increase in energy use through 2040."[56] This means that in order for the world to significantly decrease overall emissions, China and India must make large reductions.

"[China and India] combined account for half the world's total increase in energy use through 2040."[56]

—Adam Sieminski, administrator of the US Energy Information Administration.

Reduce Fossil Fuel Use

The most effective way to reduce carbon emissions is to reduce fossil fuel use. The world can do this by developing and using alternative energy

The United States and China Must Reduce Emissions

Society can reduce climate change by cutting emissions of greenhouse gases, and the place to begin is in China and the United States. This graph shows world carbon-dioxide emissions from fossil fuels for 2012. Certain countries or regions emit more than others. Even international shipping and air traffic contribute to emissions. But the graph shows that of all the sources, the United States and China produce a large percentage of worldwide emissions. Any successful reduction in emissions must include these two countries.

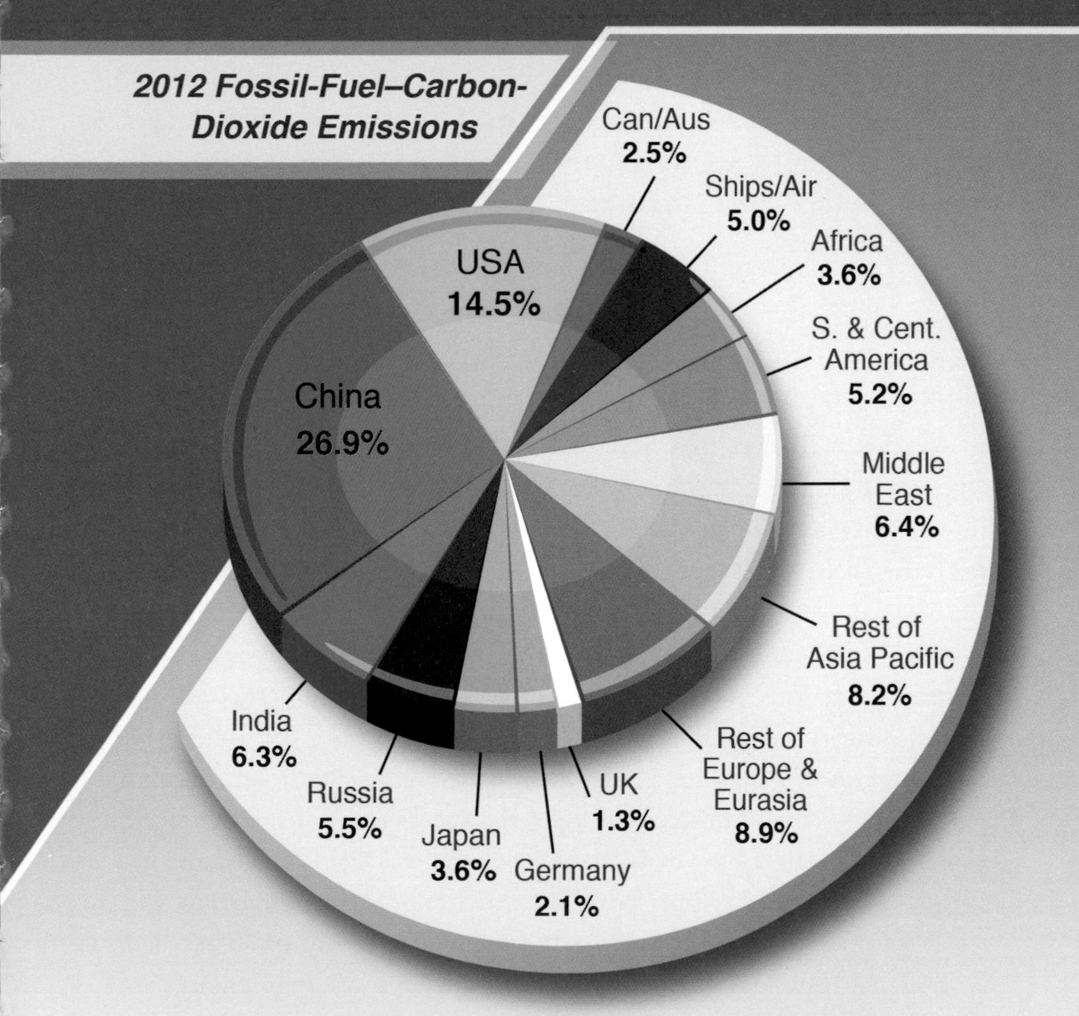

Source: James Hansen et al. "Assessing 'Dangerous Climate Change': Required Reduction of Carbon Emissions to Protect Young People, Future Generations and Nature," *PLOS One*, December 3, 2013. www.plosone.org.

sources such as solar power and wind power instead of fossil fuels. Even if fossil-fueled power remains part of the energy future, society must also embrace other options. The American Chemical Society states that almost three-quarters of the carbon dioxide generated in the United States comes from burning fossil fuels for transportation and electricity generation. It advises that instead of continuing to rely on fossil fuels, the United States should start "adopting non-combustion energy sources based on solar thermal, solar photovoltaic, wind, tidal power, or nuclear energy [which] can reduce carbon dioxide emissions produced by fossil fuel combustion for electricity generation."[57] Part of the European Union's success in reducing its emissions under the Kyoto Protocol is its use of alternative energy sources such as these. For example, it produces substantial amounts of power through the use of wind. According to the European Wind Energy Association, wind energy made up 26 percent of all new power capacity installed in 2012. The association says that wind meets 7 percent of Europe's electricity demand.

Geoengineering

Another way to reduce climate change is through geoengineering, where scientists actively manipulate Earth's climate system to stop it from changing, or reverse changes that have occurred. For example, scientists have suggested adding iron to the ocean to cause it to soak up more carbon dioxide or putting sulfate particles into the atmosphere to block some of the sun's heat. There is evidence from nature that such manipulation might be effective. For example, volcanic eruptions add large amounts of sulfate to the atmosphere and have a temporary cooling effect on the climate. Geoengineering expert Ken Caldeira explains how one of the largest eruptions in recent history cooled the climate. He says, "In 1991, a volcano in the Philippines known as Mount Pinatubo erupted and sent a huge amount of material into the stratosphere. It reflected two percent of sunlight back to space and Earth cooled by half a degree Celsius." Caldeira says the effect was only temporary but shows the potential of geoengineering. He says, "That material fell out of the atmosphere after a year or so but had that material been maintained it would have been more than enough to offset all the global warming expected this century."[58]

Critics argue that to interfere with Earth's climate is risky because there may be unpredictable negative consequences. If, however, climate change continues at the same rate, then geoengineering might be the best option. Society should work to develop geoengineering solutions now in case climate change becomes so severe that geoengineering is worth the risk. Caldeira says, "Emissions and global temperatures keep going up and up and up. We need to think about what we'll do if bad things happen."[59]

"Emissions and global temperatures keep going up and up and up. We need to think about what we'll do if bad things happen."[59]

—Ken Caldeira, geoengineering expert.

Human Action Will Make a Difference

While some climate change is inevitable, human action can still make a substantial difference to the world's future climate. In his 2013 Climate Action Plan, Obama insists that not only will human action make a difference, but action is imperative. He says, "We have a moral obligation to future generations to leave them a planet that is not polluted and damaged." He insists, "We can protect our children's health and begin to slow the effects of climate change so that we leave behind a cleaner, more stable environment."[60] The sooner the world takes action to reduce climate change, the less harm it will experience.

Society Should Not Try to Reduce Climate Change

"No plausible U.S. action changes the overall trajectory of emissions and warming or the nature of the potential impact."

—Cass was domestic policy director of politician Mitt Romney's 2012 presidential campaign.

Oren Cass, "The Next Climate Debate: Conservatives Should Accept the Science and Focus on Policy," National Review, March 25, 2013. www.nationalreview.com.

Consider these questions as you read:

1. Can you think of an example of how fossil fuels have improved human life? Explain.
2. How strong is the argument that developing nations need to pursue economic growth, even if it causes some environmental harm? Explain.
3. How persuasive is the argument that society should not try to reduce climate change? Which arguments provide the strongest support for this perspective?

Editor's note: The discussion that follows presents common arguments made in support of this perspective, reinforced by facts, quotes, and examples taken from various sources.

Some scientists explain climate change by using the analogy of a bathtub. The bathtub represents the world and its atmosphere, and as humans produce carbon dioxide and other greenhouse gases, they are gradually filling the bathtub. This tub does contain a plug—a way for carbon dioxide to escape—however, only a little seeps out at a time as the environment naturally absorbs it. Most of the carbon is left sitting in the bathtub for a long time, and because humans are producing such large amounts the tub is getting very full. As a result, climate change is occurring. In addition to showing how carbon causes warming, the bathtub analogy also illustrates the problem with trying to fix the situation: No matter what humans do, the bathtub is already very full of carbon dioxide, and because it takes so long to drain it will remain full for a long time. Henry Shue, research fellow and professor at the University of Oxford, argues that carbon dioxide stays in the atmosphere for close to a thousand years. Therefore, he insists, humans cannot reduce it no matter what policies they make. He says, "So for any practical matter, any policies or decisions that humans want to make, carbon dioxide is effectively just staying there."[61] This means that society should not take action to reduce climate change because nothing can be done about it.

Fossil Fuels Are Necessary

The bigger problem with trying to reduce carbon emissions is that it will actually be harmful to society. Carbon emissions are primarily produced through fossil fuel consumption; fossil fuels have helped countries like the United States make myriad technological advances that have vastly improved life. Other energy sources simply do not provide an adequate substitute. While there is evidence that climate change might harm society in some ways, if society stopped using fossil fuels, the harm would be far greater because it would drastically reduce the quality of life. Journalist Jonah Goldberg argues that while climate change might have some negative effects on society, everyday life for much of the world has improved because of fossil fuels. He says, "By any conceivable measure—save, arguably, outdoor temperatures—the Earth is a vastly more hospitable place for humanity." Goldberg says, "When Pilgrims came to North America, it was often described as an inhospitable wilderness. Malaria, smallpox and yellow fever decimated immigrants (not to mention untold millions of Native Americans). Backbreaking labor was the only means of subsistence for

"For any practical matter, any policies or decisions that humans want to make, carbon dioxide is effectively just staying there."[61]

—Henry Shue, research fellow and professor at the University of Oxford.

Action on Global Warming Is Not a Priority

A 2013 Pew Research Center poll on public policy priorities reveals that most Americans do not see action on global warming as a priority. According to the poll, numerous issues take precedence over global warming as an issue that Congress and the White House must address.

Percent saying each is a "top priority" for president and Congress this year

	January 2013
Strengthening economy	86%
Improving job situation	79%
Reducing budget deficit	72%
Defending against terrorism	71%
Making Social Security financially sound	70%
Improving education	70%
Making Medicare financially sound	65%
Reducing health costs	63%
Helping poor and needy	57%
Reducing crime	55%
Reforming tax system	52%
Protecting environment	52%
Dealing with energy problem	45%
Reducing influence of lobbyists	44%
Strengthening the military	41%
Dealing with moral breakdown	40%
Dealing with illegal immigration	39%
Strengthening gun laws	37%
Dealing with global trade	31%
Improving infrastructure	30%
Dealing with global warming	**28%**

Source: Pew Research Center, "Climate Change: Key Data Points from Pew Research," November 5, 2013. www.pewresearch.org.

millions of Americans for generations. Drudgery and toil—have you ever tried to churn butter?—were necessary for even the simplest pleasures." Now, he says, "Things have gotten a wee bit warmer outside. But inside they've gotten vastly more hospitable, as we live longer, eat better, have more leisure time and have fewer deadly occupations."[62]

Steven F. Hayward, senior fellow at the Pacific Research Institute and author of the *Almanac of Environmental Trends,* insists that no good substitute exists for fossil fuels and that trying to force a change would do more harm than good. He says, "Energy is not like other goods that can be substituted or done without. It has rightly been called the master resource, because it is fundamental to everything else in the economy."[63] He argues that alternative energy sources such as solar power, biofuels, or hydrogen power cost too much to be viable alternatives to fossil fuels. The most likely result of abandoning fossil fuels would be economic upheaval.

The Importance of Economic Development

For developing nations in particular, fossil fuels drive vital economic growth, and there is little chance that these nations will stop using them even under the threat of climate change. Oren Cass, who was domestic policy director of Mitt Romney's 2012 presidential campaign, says, "The developing world has billions of people to lift out of a poverty whose depths we can barely imagine; if ameliorating poverty through economic growth creates a risk of catastrophic climate change, that is a risk they will take."[64]

China is one example of a country that values economic growth over reducing climate change. China produces one of the highest levels of greenhouse gas emissions in the world, but its people have also greatly benefited from the many technological advances that produce those emissions. Derrick Morgan, vice president for domestic and economic policy at the Heritage Foundation, says, "The Chinese people have seen a nearly sixfold increase in per capita gross domestic product (GDP) from 1990 to 2011. Hundreds of millions of Chinese have been lifted from poverty." Overall, he argues that action to protect the environment cannot come at the expense of society's well-being. He insists, "It is important to remember that environmental policy must ultimately be good for people, any country's most important resource."[65]

Geoengineering Is Too Risky

Another proposed solution to the problem of climate change is geoengineering; however, this is unpredictable and too risky to use. Geoengineering involves the active manipulation of Earth's climate system in order to stop the climate from changing or to reverse changes that have already happened. One example of a proposed geoengineering solution is to inject reflective particles such as sulfur into the air in order to block the sun and cool Earth.

The problem with geoengineering is that it is unpredictable. Attempts to alter climate could have harmful and irreversible ramifications. Andrew E. Dessler warns, "Geoengineering may lead to unintended consequences that leave the world worse off."[66] In 2014 *Nature Communications* published the results of a study of geoengineering by researchers from the Helmholtz Centre for Ocean Research Kiel in Germany. The researchers modelled five potential methods of attempting to mitigate climate change including adding lime or iron filings to the oceans, reforesting deserts, and reflecting sunlight into space. They found that such actions could cause drastic and unintended changes that might be harmful. Reflecting sunlight, for instance, would likely reduce the earth's temperature but it could also substantially alter rainfall patterns. Reforestation could change wind patterns, or even reduce tree growth in other areas. In addition, the researchers found that such efforts could not be stopped after they were begun because rapid warming would occur. Overall, they report that geoengineering is a poor strategy because it could have potentially disastrous results, and such efforts would have only a minor effect on the global temperature.

> **"Adapting to an evolving climate is going to be required in every sector of society."[67]**
>
> —Richard Moss, senior staff scientist at the US Department of Energy's Pacific Northwest National Laboratory.

Adaptation Planning

Instead of working toward the impossible goal of trying to stop climate change, society should engage in adaptation planning. This means preparing for the way the climate will change and taking steps to minimize the harm that will occur. Richard Moss of the US Department of Energy's Pacific Northwest National Laboratory says, "Adapting to an evolving climate is going to be required in every sector of society, in every region of the globe. We need to get going . . . if we are going to meet the challenge."[67]

On its website the US Environmental Protection Agency suggests some things society can do to help it adapt to the coming changes. To address potential agricultural problems it suggests breeding varieties of crops that are more tolerant to drought, heat, and heavy rainfall from flooding. For the problem of rising seas, it suggests that society leave beaches and coastal wetlands undeveloped to allow the sea to rise without destroying property. To address the potential of more severe weather events, it advises the improvement of early warning systems and evacuation plans.

Overall, society has the ability to adapt to a changing climate, and it will need to do so in the future. Any attempts to reduce climate change, such as geoengineering or lowering carbon emissions, will be ineffective and harmful to society.

Source Notes

Overview: Climate Change

1. Quoted in Internet Movie Database (IMDb), *An Inconvenient Truth* (2006). www.imdb.com.
2. Environmental Protection Agency, "Climate Change Facts: Answers to Common Questions," updated September 9, 2013. www.epa.gov.
3. Climate Central, *Global Weirdness: Severe Storms, Deadly Heat Waves, Relentless Drought, Rising Seas, and the Weather of the Future*. New York: Pantheon, 2012, pp. 27–28.
4. Intergovernmental Panel on Climate Change, "Climate Change 2013: The Physical Science Basis; Summary for Policymakers," 2013. www.climatechange2013.org.
5. American Chemical Society, "Climate Change," Public Policy Statement 2010–2013. www.acs.org.
6. Climate Central. *Global Weirdness*, p. 25.

Chapter One: Is Climate Change Caused by Human Activity?

7. Rick Docksai, "Disappearing Forests; Actions to Save the World's Trees," *Futurist*, September/October 2013. www.wfs.org.
8. Intergovernmental Panel on Climate Change, "Climate Change 2013."
9. *Los Angeles Times*, "The Obamacare Exemptions That Aren't," October 5, 2013. www.latimes.com.
10. Geological Society of America, "Climate Change," GSA Position Statement, adopted October 2006, revised April 2010; March 2013. www.geosociety.org.
11. Environmental Protection Agency, "Climate Change Facts: Answers to Common Questions," updated September 9, 2013. www.epa.gov.
12. Ralph Keeling, "'Dangerous Territory': Carbon Dioxide Levels Reach Milestone," NPR, May 10, 2013. www.npr.org.

13. Quoted in Fiona Harvey and Graham Readfearn, "Big Business Funds Effort to Discredit Climate Science, Warns UN Official," *Guardian* (London), September 20, 2013. www.theguardian.com.
14. National Aeronautics and Space Administration, "Global Climate Change: Consensus," 2013. http://climate.nasa.gov.
15. E. Kirsten Peters, *The Whole Story of Climate: What Science Reveals About the Nature of Endless Change*. New York: Prometheus, 2012, pp. 9–10.
16. Peters, *The Whole Story of Climate*, p. 13.
17. Joseph Bast, "Global Warming: Not a Crisis," *Somewhat Reasonable: The Policy and Commentary Blog of the Heartland Institute*. http://heartland.org.
18. PBS Teachers, "The Climate Change Skeptic's Argument: Natural Solar Cycles or Human Activity?," www.pbs.org.
19. PBS Teachers, "The Climate Change Skeptic's Argument."
20. Willie Soon and Sebastian Lüning, "Solar Forcing of Climate," in *Climate Change Reconsidered II: Physical Science*, Nongovernmental International Panel on Climate Change, 2013. www.nipccreport.org.
21. Roy W. Spencer, "Statement to the Environment and Public Works Committee of the United States Senate," July 18, 2013. www.drroyspencer.com.
22. Marc Morano, "New Paper Finds Ice Core CO_2 Levels Lag Temperature by up to 5,000 years–Published in Climate of the Past," *Climate Depot*, November 13, 2013. www.climatedepot.com.

Chapter Two: Has the Extent of Climate Change Been Overstated?

23. Intergovernmental Panel on Climate Change, "Climate Change 2007: Working Group II: Impacts, Adaptation and Vulnerability," 2007. www.ipcc.ch.
24. Quoted in Fiona Harvey and Graham Readfearn, "Big Business Funds Effort to Discredit Climate Science, Warns UN Official."
25. Nate Cohn, "Explaining the Global Warming Hiatus: Grappling with Climate-Change Nuance in a Toxic Political Environment," *New Republic*, June 18, 2013. www.newrepublic.com.

26. Spencer, "Statement to the Environment and Public Works Committee."
27. Judith A. Curry, "Statement to the Committee on Environment and Public Works of the United States Senate," hearing on "Review of the President's Climate Action Plan," January 16, 2014. http://scienceandpublicpolicy.org.
28. Paul C. "Chip" Knappenberger, "UN'S New Climate Change Report an Embarrassment, Self-Serving and Beyond Misleading," Cato Institute, September 26, 2013. www.cato.org.
29. Steven F. Hayward, "Q&A: Steven F. Hayward on Climate Change," *Arch Conservative*, October 2, 2013. www.archconuga.com.
30. *National Geographic*, "Larsen Ice Shelf Expedition." www.nationalgeographic.com.
31. *National Geographic*, "Larsen Ice Shelf Expedition."
32. Donald R. Prothero, "How We Know Global Warming Is Real and Human Caused," *Skeptic*, vol. 17, no. 2, 2012. www.skeptic.com.
33. Michael Le Page, "Global Warming," *New Scientist*, November 11, 2012. www.newscientist.com.
34. Quoted in Tim Folger, "Rising Seas," *National Geographic*, September 2013. http://ngm.nationalgeographic.com.
35. Intergovernmental Panel on Climate Change, "Climate Change 2013."
36. Quoted in Alex Brown, "Study: Oceans Absorbing Much of Climate Change's Impact," *National Journal*, November 1, 2013. www.nationaljournal.com.
37. Quoted in Richard Harris, "A Cooler Pacific May Be Behind Recent Pause In Global Warming," NPR, August 29, 2013. www.npr.org.
38. Barack Obama, "Remarks by the President on Climate Change," Georgetown University, Washington, DC, June 25, 2013. www.whitehouse.gov.

Chapter Three: Will Climate Change Be Harmful to Society?

39. Andrew E. Dessler, *Introduction to Modern Climate Change*. New York: Cambridge University Press, 2011, p. 136.

40. Deborah Zabarenko, "2010 Russia Heat Wave Due to Natural Variability: U.S.," *Reuters*, March 9, 2011. www.reuters.com.
41. Coral Davenport, "The Scary Truth About How Much Climate Change Is Costing You," *National Journal*, February 7, 2013. www.nationaljournal.com.
42. Jeffrey Goldberg, "Drowning Kiribati," *Bloomberg Business Week*, November 21, 2013. www.businessweek.com.
43. Environmental Protection Agency, "Climate Impacts on Ecosystems," updated September 9, 2013. www.epa.gov.
44. Michael Hutchins, interview by Jordan Carlton Schaul, "The Climate Change Conundrum: What the Future Is Beginning to Look Like," *National Geographic*, January 10, 2013. www.nationalgeographic.com.
45. Bradley J. Dibble, *Comprehending the Climate Crisis: Everything You Need to Know About Global Warming and How to Stop It*. Bloomington, IN: iUniverse, 2012, p. 94.
46. Patrick J. Michaels, "Addendum: Global Climate Change in the United States," Cato Institute, October 31, 2012. www.cato.org.
47. Anne Jolis, "The Weather Isn't Getting Weirder," *Wall Street Journal*, February 10, 2011. www.wsj.com.
48. Marco Evers, Olaf Stampf, and Gerald Traufetter, "Climate Catastrophe: A Superstorm for Global Warming Research," *Spiegel Online International*, April 1, 2010. www.spiegel.de.
49. James Vlahos, "Why Some Coral Reefs Might Survive Climate Change," *Popular Science*, August 2013. www.popsci.com.
50. Claude Allegre et al. "No Need to Panic About Global Warming," *Wall Street Journal*, December 4, 2013. www.wsj.com.
51. Craig D. Idso, "The Positive Externalities of Carbon Dioxide: Estimating the Monetary Benefits of Rising Atmospheric CO_2 Concentrations on Global Food Production," Center for the Study of Carbon Dioxide and Global Change, October 21, 2013. http://co2science.org.
52. Nongovernmental International Panel on Climate Change, "Climate Change Reconsidered II: Physical Science: Chapter 7: Observations: Extreme Weather," 2013. http://climatechangereconsidered.org.
53. Quoted in Jolis, "The Weather Isn't Getting Weirder."

Chapter Four: Should Society Try to Reduce Climate Change?

54. Barack Obama, "Remarks by the President on Climate Change," Georgetown University, Washington, DC, June 25, 2013. www.whitehouse.gov.
55. Richard W. Miller, "'Global Suicide Pact': Why Don't We Take Climate Change Seriously?," *Commonweal*, March 23, 2012. www.commonwealmagazine.org.
56. Quoted in "Developing Countries' Carbon Emissions Will Vastly Outpace Developed Nations, U.S. EIA Says," *Huffington Post*, July 25, 2013. www.huffingtonpost.com.
57. American Chemical Society, "Climate Change," Public Policy Statement, 2010–2013. www.acs.org.
58. Quoted in David Biello, "What Is Geoengineering and Why Is It Considered a Climate Change Solution?," *Scientific American*, April 6, 2010. www.scientificamerican.com.
59. Quoted in Biello, "What Is Geoengineering?"
60. Executive Office of the President, "The President's Climate Action Plan," June 2013. www.whitehouse.gov.
61. Henry Shue, "Climate Change, Human Rights, and the Trillionth Ton of Carbon," in Jonathan Beever and Nicolae Morar, *Bioethics, Science, and Public Policy*. West Lafayette, IN: Purdue University Press, 2013, p. 73.
62. Jonah Goldberg, "Inhospitable Earth—Compared to What?," American Enterprise Institute, July 2, 2013. www.aei.org.
63. Steven F. Hayward, "No: Alternatives Are Simply Too Expensive," *Wall Street Journal*, September 21, 2009. http://online.wsj.com.
64. Oren Cass, "The Next Climate Debate: Conservatives Should Accept the Science and Focus on Policy," *National Review*, March 25, 2013. www.nationalreview.com.
65. Derrick Morgan, "A Carbon Tax Would Harm U.S. Competitiveness and Low-Income Americans Without Helping the Environment," Heritage Foundation Backgrounder # 2720, August 21, 2012. www.heritage.org.

66. Dessler, *Introduction to Modern Climate Change*, p. 181.
67. Quoted in Tim Radford, "Scientists Deliver Ultimatum: Adapt to Climate Change or Go Under," Responding to Climate Change, November 11, 2013. www.rtcc.org.

Climate Change Facts

Evidence of Climate Change

- According to the American Meteorological Society, all of the ten warmest years in the global temperature records up to 2011 have occurred since 1997.
- The National Aeronautics and Space Administration (NASA) reports that the continent of Antarctica has been losing more than 100 cubic kilometers of ice each year since 2002.
- The Environmental Protection Agency reports that over the last century, the average global temperature increased by more than 1.4°F.
- The *Economist* reports that the extent of Arctic sea ice has shrunk by 3.5–4.1 percent a decade between 1979 and 2012.

Climate Change and the Oceans

- NASA reports that as a result of increasing carbon dioxide emissions, the acidity of surface ocean waters has increased by about 30 percent since the beginning of the Industrial Revolution.
- According to NASA, global sea-level rise in the last decade is nearly double that of the last century.
- The Ocean Conservancy estimates that more than 50 percent of Americans live within fifty miles (80.5 km) of the coast.
- In a 2012 report by a group of experts organized by National Oceanic and Atmospheric Administration, it was estimated that by 2100 the ocean could rise as much as 6.6 feet (2 m).

Greenhouse Gas Emissions

- According to the Environmental Defense Fund in 2013, current carbon dioxide levels are 40 percent higher than the highest levels over the past eight hundred thousand years.

- Regina McCarthy, administrator of the US Environmental Protection Agency, says that motor vehicles emit nearly a third of US carbon pollution.
- The International Energy Agency reports that in 2011 the United States alone contributed 16.9 percent of global carbon dioxide emissions; only China produced more.
- According to the Nature Conservancy in 2012, 15 percent of carbon emissions come from deforestation.

Societal Impacts

- The World Health Organization estimates that climate change could expose an extra 2 billion people to dengue fever by the 2080s.
- In his 2013 Climate Action Plan, Barack Obama states that in 2012 weather and climate disaster events caused over $110 billion in estimated damages.
- The Conservation Fund reports that in the northeastern United States, the frost-free start of early spring now comes eleven days earlier than it did in the 1950s. It says that if current climate trends continue, the frost season could be shortened by a full month by 2050.
- In a 2013 report the World Bank estimates that as a result of climate change, by the 2030s 40 percent of the land now growing maize in sub-Saharan Africa will no longer be able to support the crop.

Climate Change Polls

- According to the Pew Research Center, in a 2013 poll of 37,653 people in thirty-nine countries, 54 percent believed climate change was a major threat to their countries.
- According to the NASA, 97 percent of climate scientists agree that warming trends over the past century are very likely due to human activity.

- In a 2013 poll of 1,022 adults, Gallup found that 41 percent of people believe that news reports about global warming are exaggerated.
- In a 2013 poll of 1,504 adults by the Pew Research Center, two-thirds of Americans say there is solid evidence that Earth has been getting warmer over the last few decades.

Related Organizations and Websites

American Enterprise Institute (AEI)
1150 17th St. NW
Washington, DC 20036
phone: (202) 862-5800 • fax: (202) 862-7177
website: www.aei.org

The American Enterprise Institute is a community of scholars and supporters committed to expanding liberty, increasing individual opportunity, and strengthening free enterprise. It advocates environmental policy that does not infringe on democratic institutions or human liberty. Its website includes links to numerous articles about climate change.

Center for Climate and Energy Solutions
2101 Wilson Blvd., Suite 550
Arlington, VA 22201
phone: (703) 516-4146 • fax: (703) 516-9551
e-mail: press@c2es.org
website: www.c2es.org

The Center for Climate and Energy Solutions is a nonprofit organization working to address the challenges of energy and climate change. It is committed to advancing safe, reliable, and affordable energy for the world while protecting the climate.

Center for the Study of Carbon Dioxide and Global Change
PO Box 25697
Tempe, AZ 85285-5697
phone: (480) 966-3719 • fax: (480) 966-0758
e-mail: contactus@co2science.org
website: http://co2science.org

The Center for the Study of Carbon Dioxide and Global Change was created to disseminate reports and commentary about new developments related to the consequences of rising carbon dioxide levels. It publishes the magazine *CO_2 Science*, which contains editorials and reviews of scientific journal articles.

Environmental Defense Fund
257 Park Ave. South
New York, NY 10010
phone: (800) 684-3322
website: www.edf.org

The Environmental Defense Fund's mission is to preserve the natural systems that life depends on. It believes society must take action, or climate change will have catastrophic effects. The organization's website has information about what it believes are the most effective ways to address the problem.

Heartland Institute
One South Wacker Dr., # 2740
Chicago, IL 60606
phone: (312) 377-4000 • fax: (312) 377-5000
e-mail: think@heartland.org
website: http://heartland.org

The Heartland Institute works to promote free-market solutions to social and economic problems. Its website contains numerous policy studies about climate change, which it argues is not a serious problem.

Intergovernmental Panel on Climate Change (IPCC)
C/O World Meteorological Organization
7bis Avenue de la Paix
C.P. 2300
CH-1211 Geneva 2, Switzerland
phone: +41-22-730-8208/54/84 • fax: +41-22-730-8025/13
e-mail: IPCC-Sec@wmo.int
website: www.ipcc.ch

The Intergovernmental Panel on Climate Change (IPCC) is the leading international body for the assessment of climate change. Thousands of scientists from all over the world contribute to the work of the IPCC on a voluntary basis. Its website includes reports, graphics, and speeches about climate change.

National Oceanic and Atmospheric Association (NOAA)
1401 Constitution Ave. NW, Room 5128
Washington, DC 20230
phone: (301) 713-1208
website: www.noaa.gov

NOAA researches the conditions of the oceans and the atmosphere and supplies information about atmospheric and weather conditions to the public. Its website contains numerous fact sheets and articles about climate change.

US Environmental Protection Agency (EPA)
1200 Pennsylvania Ave. NW
Washington, DC 20460
phone: (202) 272-0167
website: www.epa.gov

The US Environmental Protection Agency is the federal agency that works to protect human health and the environment. It performs research, develops environmental regulations, and educates the public about environmental issues. Its website contains numerous fact sheets about climate change.

"What We Know," American Association for the Advancement of Science (AAAS)
1200 New York Ave. NW
Washington, DC
phone: (202) 326 6400
website: http://whatweknow.aaas.org

In March 2014 the AAAS launched a website and issued an eighteen-page report that clearly and succinctly lays out many effects of human-caused climate change already underway and warns of severe consequences the longer governments delay action. The report, entitled "What We Know: The Reality, Risks and Response to Climate Change," was compiled by a panel of thirteen US climate scientists including oceanographers, ecologists, and public health experts. The scientists are members of the AAAS, which has more than 120,000 members and is the world's largest general scientific society. The website includes links to the downloadable report, to the comments of scientists, and to a five-minute video entitled "Consensus Sense."

For Further Research

Books

Climate Central, *Global Weirdness: Severe Storms, Deadly Heat Waves, Relentless Drought, Rising Seas, and the Weather of the Future*. New York: Pantheon, 2012.

Bradley J. Dibble, *Comprehending the Climate Crisis: Everything You Need to Know About Global Warming and How to Stop It*. Bloomington, IN: iUniverse, 2012.

E. Kirsten Peters, *The Whole Story of Climate: What Science Reveals About the Nature of Endless Change*. New York: Prometheus, 2012.

H.W. Wilson Company, *The Reference Shelf: Global Climate Change*. Armenia, NY: Grey House, 2013.

Periodicals

Nate Cohn, "Explaining the Global Warming Hiatus: Grappling with Climate-Change Nuance in a Toxic Political Environment," *New Republic*, June 18, 2013.

Coral Davenport, "The Scary Truth About How Much Climate Change Is Costing You," *National Journal*, February 7, 2013.

Tim Folger, "Rising Seas," *National Geographic*, September 2013.

Richard W. Miller, "'Global Suicide Pact': Why Don't We Take Climate Change Seriously?," *Commonweal*, March 23, 2012.

Donald R. Prothero, "How We Know Global Warming Is Real and Human Caused," *Skeptic*, vol. 17, no. 2, 2012.

Internet Sources

Environmental Protection Agency, "Climate Change," October 23, 2013. www.epa.gov/climatechange.

Executive Office of the President, "The President's Climate Action Plan," June 2013. www.whitehouse.gov/sites/default/files/image/president27s climateactionplan.pdf.

Intergovernmental Panel on Climate Change, "Climate Change 2013: The Physical Science Basis," 2013. www.ipcc.ch/report/ar5/wg1/#.Uu CUANrTldg.

Patrick J. Michaels, "Addendum: Global Climate Change in the United States," Cato Institute, October 31, 2012. www.cato.org/pubs/Global -Climate-Change-Impacts.pdf.

Nongovernmental Panel on Climate Change, "Climate Change Reconsidered II: Physical Science," October 17, 2013. http://climatechangere considered.org.

Index

Note: Boldface page numbers indicate illustrations.

adaptations
  are necessary, 10–11, 36, 40–41, 46–47, 49, 60–61
  have succeeded in past, 44–46
Aegean civilizations, 37
agriculture
  adaptations are answer, 61
  historic droughts caused by climate change, 39
  increase in carbon dioxide benefits, 36, **45**, 46–47
Almanac of Environmental Trends (Hayward), 30, 59
alternative energy sources
  are necessary, 52, 54
  cannot replace fossil fuels, 57, 59
American Chemical Society, 10–11, 54
American Meteorological Society, 33–34
Antarctica
  ice core samples, 16–17, 23
  ice shelf is melting, **8**, 31–32, 33
Arctic ice masses, melting of, 32–33, **34**
automobiles, 14

Bast, Joseph, 21, 25
bathtub analogy, 56–57

Caldeira, Ken, 54, 55
carbon dioxide emissions, 7
  by country, **53**
  geoengineering can reduce, 54
  increase in
    benefits agriculture and all plants, 36, **45**, 46–47
    does not correlate with warming, 23
    is caused by human activity, 12, 13–14, **15**, 16–17
  must be reduced, 51, 52
  from power plants, 51
Cass, Oren, 56, 59
Cato Institute, 44
Center for the Study of Carbon Dioxide and Global Change, 47
China, 52, **53**, 59
civilizations, collapse of, 37–38
climate, factors in determining, 6–7, 8–9
Climate Action Plan (Obama, 2013), 55
Climate Central, 9, 11
clouds, effects of, 9
Cohn, Nate, 27
coral reefs, 46
Curry, Judith A., 28

Davenport, Coral, 39
deforestation, 12, 13–14
Deschenes, Olivier, 46
Dessler, Andrew E., 31, 39, 60
Dibble, Bradley J., 42
disease, increase in spread of, 36, 41–42
Docksai, Rick, 14
droughts, ancient, 38

Ebell, Myron, 26
ecosystems
  balances in, will be upset, 40–41
  effect of changes in, 34, 36
  interconnectedness of, 9, 41
  will adapt successfully, 36, 46–47
Egyptians, ancient, 37
Ehrlich, Anne, 41
Ehrlich, Paul, 41
energy, receipt and release of, 6–7
Environmental Defense Fund (EDF), 13
Environmental Protection Agency (EPA), 7, 16, 41, 61
EU-15, 50–51
European Union, 50–51, 54
European Wind Energy Association, 54
extent of climate change
  has been overstated, 24
    has already stopped, 25

predictions have been exaggerations, 26–27
has not been overstated, 24
global melting of ice masses proves, 31–33
temperature increases, 33–34, 35

feedback effects, 9
floods, 28
Food and Agriculture Organization (FAO), 13
fossil fuel use
causes greenhouse gases, 9, 16
drives economic growth, 59
is necessary, 57, 59
must be reduced, 49, 52, 54
oil and coal industries earnings and, 17–18
reducing will harm society, 49
in United States, 51, **53**, 54
See also carbon dioxide emissions

geoengineering, 49, 54–55, 60
Geological Society of America (GSA), 16
geologic record, 15, 21
glaciers
shrinking of, 24, 25, 32
water contained in, 26
God, Creation, and Climate Change (Miller), 51–52
Goldberg, Jeffrey, 40
Goldberg, Jonah, 57, 59
Gore, Al, 6
greenhouse effect, 7–9
greenhouse gases
cause climate change, 7–9
cause extreme weather events, 39
causes of emissions, 14
do not influence climate, 21
do not cause temperatures to rise, 12, 22–23
efforts to reduce emissions, 10
See also carbon dioxide emissions
Greenland, 17, 32, 33

Hansen, James, 37
Hawaii, 46
Hayward, Steven F., 30, 59
heat waves, 39
Helmholtz Centre for Ocean Research (Germany), 60
Himalayas, 25–26
Hittites, 37
Horton, Radley, 33
human activity
causes climate change, 12, 16
carbon dioxide balance, **15**, 16–17
deforestation, 13–14
greenhouse gases, 8–9
scientific consensus on, 14, 18
does not cause climate change, 12
business interests of believers, 17–18
greenhouse gases levels do not correlate with warming, 22–23
is natural process, 12
solar energy variations do, 22
variations are historically normal, 19–22, **20**, **29**
preparation for different climate is necessary, 10–11
technological adaptations have been successful, 43–45
Hutchins, Michael, 41

ice masses, melting of
in Antarctica, **8**, 31–32, 33
in Arctic, 32–33, **34**
effects of, 26, 31
in Greenland, 32, 33
in Himalayas, 25–26
is rapid, 24
Inconvenient Truth, An (film), 6
India, 52
Intergovernmental Panel on Climate Change (IPCC)
on evidence of climate change, 10
on Himalayan glaciers melting, 25
on human activity as cause of climate change, 14
on increases in temperatures, 33
mission of, 10
models have been incorrect, 26, 28–30
International Energy Agency, 9, 52

Jolis, Anne, 44–45
Jury, Christopher, 46

Keeling, Ralph, 17
Kiribati, Republic of, 40
Knappenberger, Paul C. "Chip," 28–30
Kyoto Protocol (2005), 10, 50–51, 52

Larsen B Ice Shelf (Antarctica), **8**, 32
Le Page, Michael, 32
Los Angeles Times (newspaper), 14

Mauna Loa Observatory, 17
McCarthy, Regina, 50
methane, 7
Miller, Richard W., 51–52
Morano, Marc, 23
Moretti, Enrico, 46
Morgan, Derrick, 59
mosquitos, 42
Moss, Richard, 61
Mount Pinatubo, Philippines, 54

National Aeronautics and Space Administration (NASA), 14, 18
National Geographic (magazine), 31
natural processes, as cause of climate change, 12, 15–16
natural resources, availability of, 46
Nature Climate Change (journal), 40–41
Nature Communications (journal), 60
Netherlands, 43–44
nitrous oxide, 7
Nongovernmental International Panel on Climate Change, 22, 43, 48

Obama, Barack, 35, 51, 55
oceans, role in climate change of, 16, 35

Peters, E. Kirsten, 19–20, 21
Philippines, 54
Pielke Jr., Roger, 28, 48
plants
  will adapt or face extinction, 40–41
  will adapt successfully, 36, 46–47
  will lose habitats, 34, 36
power plants, emissions from, 51
Prothero, Donald R., 32
public opinion, 14, **58**

Rosenthal, Yair, 35
Russia, 39

Sahara Desert, 19–20
Science (magazine), 35
Scripps Institution of Oceanography, 17
sea levels, rise in, 24
  extent of, 33
  number of people impacted by, **38**
  predicted, 6
  society has adapted before, 43–44
  suggested adaptations to, 61
  threaten coastal communities, 36, 40
Sieminski, Adam, 52
skeptics, business interests of, 17–18
snow patterns, 33–34
society
  adaptation is answer, 49
  should not try to reduce climate change, 49
    fossil fuels are necessary, 57, 59
    will not make difference, 56–57
  should try to reduce climate change, 49
    fossil fuel use should be reduced, 52, **53**, 54
    by geoengineering, 54–55
    international efforts have helped, 50–51
    US action is important, 51–52
  will be harmed, 36
    by extreme weather events, 39
    has happened historically, 37–38
    increase in disease, 36, 41–42
    as members of changing ecosystems, 41
    by rising sea levels, 36, **38**, 40
  will not be harmed, 36
    adaptations have been successful, 43–45
    ecosystems will adapt, 46–47
    weather extremes have not increased, 24, 27–28, 47–48
solar energy variations and climate, 6–7, 22
Solomon, Susan, 35
species loss, effect of, 41
Spiegel (magazine), 46
Spencer, Roy W., 23, 27
sunspots, 22

Taylor, James M., 21
temperatures
  are evidence of warming, 7, **8**
  effects of increase in, 33–34
  greenhouse gases do not cause rise in, 12, 22–23
  historically have varied, 19–22, **20**, **29**
  pause in increase in, is temporary, 35
  predictions about rise in, have been exaggerated, 27
  warm, are less threat to humans than cold, 46
Thorgeirsson, Halldór, 18
Tomes, Ray, 19

United Nations, 13
  See also Intergovernmental Panel on Climate Change (IPCC)
United States
  as chief obstacle to preventing climate change, 51–52, **53**
  cost of extreme weather events in, 39
  EPA, 7, 16, 41, 61
  fossil fuel use in, 51, **53**, 54
  Kyoto Protocol and, 10
  NASA, 14, 18
  population living on shoreline, 40
  snow patterns in, 33–34

vector-borne diseases, 41–42
Vlahos, James, 46
volcanic eruptions, 8, 35, 54

Wall Street Journal (newspaper), 47
Warren, Rachel, 40–41
weather amnesia, 28
weather extremes
  are costly, 39
  are not caused by climate change, 36
  have increased, 7, 36, 39
  have not increased, 24, 27–28, 47–48
  prediction of increase in, 6
  predictions cannot be made, 48
  suggested adaptations to, 61
wildlife
  will adapt or face extinction, 40–41
  will adapt successfully, 36, 46–47
  will lose habitats, 34, 36
wind power, 54
World Bank, 40
World Health Organization, 14, 41–42
World Resources Institute, 14
World Wildlife Federation, 13

Zabarenko, Deborah, 39